TEN JU

Birds, Beauty and Meditations in the Neotropics

NIC KORTE

outskirts press

Ten Jungle Days
Birds, Beauty and Meditations in the Neotropics

v6.0

Outskirts Press, Inc.
http://www.outskirtspress.com

ISBN: 978-1-9772-2541-2

PRINTED IN THE UNITED STATES OF AMERICA

TABLE OF CONTENTS

PREFACE i

URBAN JUNGLE 1

WHY AM I HERE? 19

IMMERSION 45

PATHWAYS 65

WINNING THE FINAL FOUR 85

BLESSED BY LIGHT 116

SUFFICIENT NOT SATISFYING 150

YOU DON'T BELONG HERE 168

RECONCILIATION 181

I DON'T WANT TO GO! 194

ACKNOWLEDGEMENTS 205

For Zia and Koen

PREFACE

> *To be read. To be heard. To be seen. I want to be read. I want to be heard. I don't need to be seen. To write requires an ego, a belief that what you say matters. Writing also requires an aching curiosity leading you to discover, uncover, what is gnawing at your bones.*
>
> — *Terry Tempest Williams,* When Women Were Birds: Fifty-four Variations on Voice

I have always been a reader. I have always enjoyed writing. I have hundreds of pages of journals describing my thoughts, reactions to books, and most of my travels. Those journals were a source for much of what is found within these pages.

My professional career was sustained by writing. I worked alone for nearly two decades as a geochemical consultant. Mostly, I evaluated complex datasets obtained to identify contaminants in the environment. While my data interpretation skills were strong, my ability to "write-up" the data was the principal reason my consulting career was successful.

Besides my technical writing, which included a book on environmental data interpretation, I've written a column for the local Audubon Society's newsletter for decades. For five years, I wrote a natural history blog for the local newspaper. I published a few articles in natural history journals appearing as long ago as the mid-1980s.

Besides being born a reader and a writer, I was born with a passion for nature, particularly for birds. A young Swiss lady, a tripmate on a naturalist journey in the Galapagos, but decidedly not a birder, observed my birding passion and asked, "What is about birds?" Her thick accent accentuated the fact that she was dumbfounded. I believe if she'd asked honestly, she would have said, "What's the matter with you?" How does one explain innate tendencies? I always wanted to know the names of plants and animals. When I learned what a pantheist was in high school, I thought, "I'm one of those!"

When I was six years old, my parents bought a set of *World Book Encyclopedias* and said they *were for me*. I took that to heart and paged though them frequently—usually to look at the photos. Mom for her part, was good at saying, "look it up" when I had a question. I have saved the "B" volume from those ancient encyclopedias. When the pages are viewed from the side, it is apparent that some are very worn—the entry on birds had many photos. Thus, unlike many who became interested in birding as teens or adults, I was always interested. Often, one hears of the child who wanted to take everything apart to see what it was made of. Well, I was that way about nature. I was born wanting to know "what is it, and why does it look or behave as it does?"

I was six or seven years old when I made my first identification. As I walked home from school, I was sure I had seen a cockatoo. My Mom laughed at me. Now I know it was probably a female Northern Cardinal. At least I had the coloration correct. In those days, I would hunt for and check out books about birds from the public library.

I've always kept a pair of binoculars close by as an adult. I also think anyone who knows me well would agree that once I'm interested in something, I usually want to learn more and more. I once read that birders, in general, were much better versed regarding natural systems than hunters. Indeed, birding and hunting have much in common except a birder takes the field with hundreds of species on his mind, not just one or two.

A relative once told me that although he couldn't understand it, he envied my enjoyment of birding because I always had something to do. That's true. Everywhere I go, I'm curious about what lives there. What makes the species distribution at one location different than another? Why are the species we are seeing in the same location different from one year or one season to another? My curiosity about such questions never stops.

An early inspiration was Henry David Thoreau. My high school classmates complained about reading some excerpts, but I devoured *Walden* and soon owned several other volumes of his work.

During and just after my college years, I encountered Joseph Wood Krutch, and Aldo Leopold. The latter's *A Sand County Almanac* would probably be my natural history selection, if I had to pick a single work as the greatest of all time. Leopold's description of sitting at The Shack while recording the time various birds sang was an inspiration. Joseph Wood Krutch is better known for *The Voice of the Desert,* but my favorite is *A Desert*

Year, which I have probably read ten or more times. His juxtaposition of life lessons with natural history observations captivated me. I didn't consciously emulate anyone's style for this book, but Krutch's *A Desert Year* may be the closest example.

I urge everyone to seek out natural history classics by the authors I have already mentioned and include others such as Sigurd Olson, Annie Dillard, Terry Tempest Williams, Ann Zwinger and more. These works are a window into the natural world both "out there" and inside one's self. I discover something new every time I re-read these authors. Sadly, I also find that these books are, more and more, living up to the "history" part of natural history, as our planet continues to lose wild places. These books should be read so that we don't forget how it was, and so we can maintain inspiration to preserve what remains. My hope is that this book, my book, can do the same.

Listening at Leopold's Shack.

CHAPTER 1

URBAN JUNGLE

He prayeth best, who loveth best
All things both great and small.

— *Samuel Taylor Coleridge,*
"The Rime of the Ancient Mariner"

"Hoop-hoop.......hoop-hoop........hoop-hoop." I peered carefully through the vegetation. The bird was sitting quietly inside the deep shade—a Lesson's Motmot. It swished its long tennis racket-like twin tails. No one has explained why the tails of these birds have a long naked sheath with a paddle-like tip. There with the garbage floating by and the putrid smell of the polluted creek sat this gaudy bird with the electric blue crown.

The church a few blocks away is inscribed 1797. I suppose this creek has been abused for centuries, but the motmot hangs on. My plane had touched down in the previous evening's twilight and, as ever when I land in the tropics, my mind surveyed the question, *How was it before?* The lights and roads and crowds of Costa Rica's populous Valle Central spread in all directions. This was all jungle once. It is all jungle now, but a different type.

My friends Mario and Raquel are all smiles and waves as they greet me at the airport. Into the car, pay the toll, drive the newly repaired road past the creek/ditch/sewer, where I'm now seeing the motmot amid the trash, then, a couple of blocks farther to their small but beautiful and modern home.

I had awakened to a variety of bird calls, including a rooster out there somewhere. This is the first day of a ten-day trip—one-day here in Heredia with my friends and nine at a remote jungle lodge. Always, every morning, the first thing I do is walk out in their yard and take a deep breath so I can feel my new climate. I stand on tiptoes and peer over the razor-wire-covered walls at the mountains that ring the Central Valley. Almost always, they are shrouded with clouds. In my imagination, I feel those mists and hear distantly calling bellbirds. I wish it were so. The mists are far away, and maybe, the bellbirds farther still. So much is now agriculture and development. Still, every time I return, I love the homecoming and am excited at the prospect of new sightings and new experiences.

Mario and Raquel's yard is tiny by US standards but a source of great pride for Mario as he tends several varieties of orchids and embarks, it seems, on continuous change. One year there are chayotes, another year oranges, then only flowers (always beautiful orchids) and then oranges again.

Before we became friends, Mario's interest in birds didn't extend much beyond *pollo frito* (fried chicken), but now he notices what comes to his yard—mostly *"comemaíz"* (eats corn) along with *yigüirros* (a word that my attempts to say usually cause laughter—it is something like "whee-gweerrrro" with well-rolled r's that do not come naturally to older English speakers). We know the *yigüirro* as the Clay-colored Thrush—a very close relative of the American Robin. Most travelers initially find it strange that a country boasting almost 900 species of birds, many very colorful, has chosen this plain-brown species as its national bird. It makes sense to the ear. Most North Americans would recognize the friendly song as robin-like and cheerful. Apparently, early coffee farmers linked its song to the blooming of coffee plants.

The most common yard bird in the Central Valley, however, is the little *comemaíz* or Rufous-collared Sparrow. These are handsome birds with a perky, little crest. They tunefully emit a sweet little trill that sounds something like *teeooo, teeeee.* I can't help but compare these with yard "sparrows" in the US—the non-native, inaptly named "House" or "English" Sparrows which aren't sparrows at all. These quarrelsome invaders shouldn't even be mentioned in the same sentence as the friendly "rufi."

Fortunately, House Sparrows have not taken over Costa Rica, but they are present. Their "natural" habitat seems to be gas stations. When I have been on long birding trips, we usually only add them to our day's list while stopped for gas. Apparently, Rufous-collared Sparrows are able to withstand the House Sparrow invasion, and are, perhaps, a bit more adaptable to humans than the native sparrows of the US. Rufous-collared Sparrows and Clay-colored Thrushes maintain some of the countryside in Costa Rica's urban yards.

My list for Mario and Raquel's tiny yard, barely four by four meters, includes Blue-gray and Palm Tanagers, Tropical Kingbirds, migrant Wilson's and Yellow Warblers,

Buff-throated and Grayish Saltators, Great Kiskadee, Social Flycatcher, and Crimson-fronted Parakeets. I can't help but contrast this with my own much larger backyard in Colorado, which I've carefully crafted to attract birds and which includes a variety of bird feeders. Two of my most common birds are House Sparrows and Eurasian-collared Doves, neither of which are native.

I often wonder what my yard would be like without the exotic species that have taken over so much of North America. If there were no House Sparrows, would a Song Sparrow live under my snowberry bushes, or, perhaps, a towhee abide in my mini orchard? I can't say, but it is enjoyable to see the native birds that hang on amid this tropical city. Even if you walk a few blocks to the crowded central square, amidst the hordes of pigeons, you can find Great Kiskadees, Tropical Kingbirds, Hoffman's Woodpeckers, and Crimson-fronted Parakeets. I even found a Boat-billed Flycatcher in the park one day. A park in a nearby small city, for many years, had been known for a pair of Black-and-white Owls. If you wanted to see them, you would look for the whitewash near the bandstand—or ask the man who sold snow cones.

But this is an incomplete view after all. This is also a country of traffic jams, noise, air pollution and crime. And, it isn't improving. Any overland drug traffic originating in South America must pass through as well. Laundering of drug money is a major distortion to their fragile economy. This was an area of verdant jungle once, and, from the remnants that remain, it isn't difficult to imagine how diverse and spectacular it must have been.

Such thinking is not productive, nor is it fair. I typed some of this at my parent's home in Illinois, looking out at a bird feeder full of House Sparrows where probably once roamed bison by the thousands along with Prairie Chickens and Sandhill Cranes and many other species now extirpated. We have eliminated a lot more of our prairie (99%) than Costa Rica has eliminated of its jungle. This tiny country, representing 0.25% of the world's land mass and hosting 5% of the planet's biodiversity, maintains more than 25% of its land in reserves. Imagine if the US had preserved the same relative amount of prairie.

At 1130 meters (3700 feet), Costa Rica's Central Valley has a beautiful climate (daily high temperatures usually in the mid-70s). The moderate weather and fertile soils attracted agricultural pioneers. Eventually, urbanization followed. The current population of the Central Valley is approaching four million.

Besides the warblers, other neotropical migrants have passed through Mario and Raquel's yard, such as a full-plumaged male Summer Tanager. No wonder they noticed it. Cornell's "All About Bird's Website" notes that it is the only "all red bird" in North America, describing its coloration as "strawberry red." I was happy they noticed the bird and were excited to tell me about it.

I was also saddened because they said it was a "rare visitor from North America." I perused some maps and confirmed that, as has been my experience, Summer Tanagers are abundant in Costa Rica from about mid-September through mid-April. One guidebook notes they can be found "in almost any habitat that has trees." I was nonplussed to read in *The Birds of Costa Rica* that the Summer Tanager is a "very common NA [North American] migrant." It lives in Costa Rica 7 ½ months. Wouldn't I be considered "resident," if I lived in a location more than half of the year? I like to say that such neotropical migrants "live" in the tropics, but always include an annual romantic vacation to the cooler north.

Mario and Raquel's lack of familiarity with a common bird wasn't surprising. Later, during this trip, I was talking to a young, local guide. He had found a roost of a family of Sunbitterns. Sunbitterns are rare. They are the only species of their genus on the planet. Moreover, when they spread their wings, they show beautiful bright spots in red, yellow and black, making them one of the most eye-popping species in the world. I asked if he'd been showing them to people. He said, "No, Ticos don't care about Sunbitterns."

It isn't fair to single out Costa Ricans or Ticos, as they call themselves. I find a similar lack of knowledge and interest in North Americans. Still, Andres' comment reminded me of many similar experiences I've had in Costa Rica. Once, my wife and, at the time, young children visited the famous shopping area in Sarchí, and someone came running out with two very young Toucans. We have the photo. Both Ann and Adam have a baby Toucan and a happy smile on their young faces. I'm sure I was scowling. Caged native species, especially parrots, remain common, and just two years before, Mario and Raquel's neighborhood resounded with the chiming call of a Black-faced Solitaire. This call is the signature sound of the cloud forest, evoking a sense of deep green and heavy mists. Someone had one, illegally caged.

Even worse than the fact that young birds are captured is the means of doing it. Parrots and toucans are cavity nesters. The primary method of capture is to cut down the nest tree. In a country with dwindling forest, and a dwindling population of dead old trees with cavities; cutting down such a tree doesn't just remove one generation, it eliminates nesting habitat for generations to come.

The good news is that Latin Americans are learning. I was recently in Colombia where the endangered Yellow-eared Parrot (population ~1000) is barely surviving. The bird nests in Wax Palms, Colombia's national tree. Unfortunately, the Catholic Church had encouraged cutting of Wax Palms for Palm Sunday. Cutting the tallest trees was especially encouraged because those palms were "closest to heaven." These tallest and oldest trees also were most likely to have nest cavities. The population of Yellow-eared parrots was, perhaps, less than 100 as recently as 1999. Fortunately, church spokesmen have now discouraged the use of Wax Palms, and there is an active program for placing nest boxes on the palms that remain.

I recall how in 1991, our Costa Rican friends had excitedly prepared a traditional Easter dish for us. It was delicious. It contained "heart-of-palm" or palmito. In those days, it was believed that "wild palmito," harvested from the forest, tasted better than the plantation-grown variety. Our friends proudly informed us they had found a wild source. The dish didn't taste so good to me once I knew the palmito's origin. Fortunately, in 1996, Costa Rica passed a strong, conservationist Forestry Law. Its terms are so good, they are worth listing here: Forests within national reserves or on State Property are patrimony of the State and harvesting them is prohibited. Converting forests on private land to other uses is also prohibited (with certain limited exceptions via permit). Indigenous communities may only make use of their forests for subsistence. Harvesting of wood from private forests is only allowed if there is a management plan in place certified by a forestry engineer that is a member of the College of Agricultural Engineers. Harvested wood requires a permit for transit and may not be sold abroad.

> Article 3(k) recognizes four environmental services: (i) mitigation of greenhouse emissions through emissions reduction and carbon fixation, capture, storage or absorption; (ii) protection of water for urban, rural or hydroelectric use; (iii) biodiversity conservation for conservation, sustainable use, scientific investigation or genetic enhancement; and (iv) protection of ecosystems or scenic natural beauty for tourism or science.

Many people tout the United States as a world leader. Here is an area where we are not. Forestry laws with these goals would do much to bolster the rapidly weakening North American ecosystems.

Costa Rica's entry in Wikipedia amplifies the country's "green" reputation:

> Costa Rica is known for its progressive environmental policies, being the only country to meet all five criteria established to measure environmental sustainability. It is ranked fifth in the world, and first among the Americas, in the 2012 Environmental Performance Index. It was twice ranked the best performing

> country in the New Economics Foundation's (NEF) Happy Planet Index, which measures environmental sustainability, and identified by the NEF as the greenest country in the world in 2009. In 2007, the Costa Rican government announced plans for Costa Rica to become the first carbon-neutral country by 2021. In 2012, Costa Rica became the first country in the Americas to ban recreational hunting.

Unfortunately, reality is not always consistent with the excellent laws and the hype of a green rating. One of our favorite places to visit is Playa Zancudo near the border with Panama. In my explanation of why I like to stay there, I often say, "Because it isn't possible to buy a t-shirt with the beach's name on it." (Zancudo is Spanish for mosquito, so maybe that's part of it.) It is at the end of a bad road. There aren't many tourist facilities. Those that are here are mostly low-end. We stay in a converted banana worker's cabin; small, but clean and comfortable.

We arrived by accident in 2005. Two birding lodges, not far away, were full for two overlapping days in the middle of our trip. A quick internet search found this small beach area in between. Even though most of what we had heard about this part of the country had not been positive, we thought, *How bad can it be? We can stand two nights almost anywhere*. The only place that came up on our internet search was also full, but they suggested some neighboring cabins—which suit us much better. I wrote some of this book on our sixth visit there, that time for twelve days.

Now back to the forestry laws. Playa Zancudo is not known for birding, but the list for the area now exceeds 200. Habitats include beach, mangroves and some remnant forest. A species list slightly exceeding 200 is paltry for a Costa Rican location but would be a well-known hotspot in the United States, especially because the area encompassed is only four to five miles in length—and less in width. Over repeated visits, I've became the self-appointed expert on Zancudo birdlife. I was especially proud of having found Yellow-billed Cotingas. This endangered, endemic species is a mangrove specialist. The best place to view them is at a bridge on the Rincon River about three hours away by car.

The Rincon cotingas are regular. Be there at 7 AM, and you will see them. In Zancudo, I found the bird also about 7 AM, seemingly headed up the estuary, just as at Rincon. I had found them two years in a row, and, the last year, two days in a row. Same time. Same place. The birds were passing through a remnant forest patch with tall strangler figs and a thick native understory. I told a few people of my discovery. Maybe I had found another regular flight of Yellow-billed Cotingas. I couldn't wait to check again on a subsequent visit.

Disappointment was instant. A few of the fig trees still stood, but everything else was burned. Flames were still visible in the understory. Nearly all the trees had been cut. I learned the land belonged to an absentee owner who lived in the US and paid no attention. The fires and cutting and the plastic and metal shanties were the work of squatters, and there seemed to be nothing to be done about it.

I lamented the loss of my special patch to a long-time Costa Rican birding guide. He let loose: "Costa Rica in general is going to hell in a hand basket...this is happening all over and no one is stopping it or really can stop it. Everywhere I go it is getting worse...not better. It makes me want to cry at times... whether it is coffee, sugarcane, damn pineapple, heart of palm, freakin' teak plantations. My fear is in the near future Costa Rica will lose its eco-friendly status, and people will opt for other locations to visit. It would serve the Ticos right if it happens. They are living on their reputation of the past as being conservationists, and right now I don't see how anyone could assume that."

Those are some strong words from a person who once described me as "someone who loves Costa Rica as much as I do." Like my friend, I deplore pineapple plantations. I describe pineapple cultivation this way: First, you strip the land of every remnant of vegetation. Then you lay plastic pipe for pumping chemicals to the pineapple plants. It is truly hideous in terms of its environmental impact. Oil palm is almost as bad. People need jobs and people need places to live. Critics of these operations claim the profits all go to multi-national corporations and that oil palm and pineapple provide fewer jobs than traditional agriculture. I wish I had an answer.

Nonetheless, Costa Rica is a nation of very fortunate topography. The high wet mountains that traverse the country prevented the formation of a single, large indigenous population, which could have been enslaved by early colonists as occurred in so much of Latin America. Similarly, the topography also made it difficult for large plantations to thrive—at least at first. Eventually, international trade in coffee and bananas resulted in "coffee barons" and domination by foreign companies, but never to the extent of Costa Rica's neighbors.

Costa Rica also lacks gold and silver when compared to its neighbors. This lack of resources made Costa Rica into a poor, isolated, and sparsely inhabited region within the Spanish Empire. It was described as "the poorest and most miserable Spanish colony in all America" by a Spanish governor in 1719. There is great irony in this because Costa Rica is now the most favored country in Latin America by almost any index of development, economy, social services, or environment.

Of course, Costa Rican history did not escape from US influence. The major airport is named after Juan Santamaría. He must have been a general or a president, right? No, he was a teen-aged soldier. A group of US mercenaries, led by William Walker overthrew the government of Nicaragua in 1855 and attempted to conquer the other nations in Central America, including Costa Rica, in order to form a private slave-holding empire. Santamaría, a drummer boy in the Army, was the illegitimate son of a single mother. He participated in the defeat of Walker's troops at Santa Rosa, Guanacaste, at a location now known as La Casona. Walker and his soldiers retreated to Nicaragua, where in April 1856, occurred what is known as the Second Battle of Rivas. The Costa Ricans were unable to advance on Walker and his men because the building they occupied gave them a strategic advantage. On April 11, Salvadoran General José María Cañas proposed that a soldier set the building on fire with a torch. Some soldiers tried and failed. Ultimately, Santamaría volunteered on the condition that, if he were killed, someone would take care of his mother. Santamaria succeeded in setting fire to the building, which led to a decisive Costa Rican victory. He was, however, fatally wounded in the action.

Two decades later, in 1876, the governor of Heredia province decided a fort was needed. This fort is just down the street from Mario and Raquel's house, so we see it often. Apropos with the non-warlike nature of Costa Ricans, the fort's gun turrets were constructed backwards (accidentally), such that they would collect bullets rather than shield and provide flexibility of aim for rifleman inside. Fortunately, it was never used in battle and now provides an interesting way for Ticos to illustrate their natural aversion to war and their pride in not having an army.

Heredia's Old Fort

Costa Rica's peaceful nature may also be a result of fortunate timing. As in the rest of Latin America, there was a relatively recent revolution. It occurred in 1948, the same year President

Truman signed the Marshall Plan into law. José Figueres Ferrer led an armed uprising in the wake of a disputed presidential election between the previous president Rafael Angel Calderón Guardia (he served as president between 1940 and 1944) and Otilio Ulate Blanco. With more than 2,000 dead, the resulting 44-day Costa Rican Civil War was the bloodiest event in Costa Rica during the 20th century. Costa Rica's "victorious rebels formed a government junta that abolished the military altogether and oversaw the drafting of a new constitution by a democratically elected assembly. Having enacted these reforms, the junta relinquished its power, on November 8, 1949, to the new democratic government."

After the coup d'état, Figueres became a national hero, winning the country's first democratic election under the new constitution in 1953. Since then, Costa Rica has held 15 presidential elections, the latest in 2018. All of them have been widely regarded by the international community as peaceful and transparent. With uninterrupted democracy dating back to at least 1948, the country is the region's most stable.

Perhaps because the United States was preoccupied with Europe, no role was taken in the Costa Rican conflict. I have always wondered what would have happened if the US had chosen sides as it did in every other Latin American country. Witness what happened in Guatemala in 1954, when the US government, at the behest of multinational corporations, assisted in the overthrow of a peaceful and democratic election. Decades of strife followed.

We knew most of this history simply because of personal interest in the region. Our knowledge pleased Raquel, whom we met in 1988 through a teacher/exchange training program in our hometown in Colorado. My wife Mary and I had already been considering a trip to Costa Rica, being intrigued by what we were reading about the country's biodiversity and safety. What luck for all of us!

Raquel encouraged us to visit and to bring our children, because she had nieces and nephews the same age. In early 1989 we accepted her invitation. A large group met us at the airport, threw confetti on us, and whisked us away—our children (then aged 8 and 10) were in cars with people they had never met and couldn't understand. But we could all communicate through gestures and expressions—mostly smiles and laughs. I can't overstate how our wonderful experiences with Raquel and her family have made us think fondly of and feel comfortable with the American tropics.

In 1989, that first time my family and I visited Costa Rica, there were two flights per day. One arrived from Miami and one from Houston. Now, many airlines have two or more flights per day. Juan Santamaría Airport in the central valley has been extensively remodeled. The increasing urbanization is further demonstrated by the construction of a second international airport located in the northern part of the country.

Thus, the effects of the urban jungle are everywhere. Cars are parked behind bars. High walls around houses are lined with glass and razor wire. Everyone warned us about crime. "Don't roll your windows down at stoplights. Someone will reach in and steal your watch, necklace or purse." Approximately the fifth time we arrived, we were, as usual, renting a car. In conversation with the rental agent, I mentioned how frequently we had visited. "How often have you been robbed?" was his rejoinder. Well, our answer was "never," and it still is after nearly thirty visits. We have been careful, and we have been mostly in the countryside, not in the cities. So far, so good.

The sad fact is that the urban jungle continues its encroachment on the real jungle. The 2016 population was approximately 5 million, up from 4.58 million at the 2011 census, and from about 3 million on our first visit in 1989. I wonder too whether it is fair for me to regard the encroaching urban jungle as a "sad fact." Certainly, from the standpoint of biodiversity it is a disaster, but the standard of living of the average Tico has dramatically increased. We can see it in the new homes and vehicles of the younger generation of our Costa Rican family—the kids who threw confetti at us in the 1980s. Now well-educated professionals, they are aspiring and achieving a standard of living similar to the US. I don't have the right to deny them their financial success, although I still must lament the loss of natural habitat.

Emergence from the "third world" has led to an increase in vehicle ownership from approximately 140 per 1000 people in 2000 to 287 in 2017 ("All That Glitters Is Not Green," *The Guardian*). What an impact that has caused! As much as anyone, I have taken advantage of the improved roads so that I can move quickly about the country. The once multi-hour trip from San José to the Pacific Coast can now be made easily in less than an hour, depending on traffic.

Still, all the development and improved roads are painful for me to see. Recall my previous account of Playa Zancudo. The final 40 kilometers of that drive formerly required about two jolting hours of potholes and dust—or mud, depending on the season. We both complained about and celebrated the road, knowing that it contributed to the area's isolation. An added benefit were many small wetland areas that were well worth a brief stop. One was a favorite. The area was perhaps 60 or 70 meters in diameter on one end of a pasture. On previous visits, we had found the small marsh replete with Northern Jacanas, Bare-Throated Tiger-Herons, three species of egrets and more. Adjacent to the marsh was a small copse of five or six trees—not much bigger than shrubs. These trees always held an amazing array of birds: saltators, warblers, and tanagers. A few flowering bushes nearby had once yielded a Veraguan Mango—a rare hummingbird.

On our 2016 visit, we once again planned a brief stop at this small marsh. Somehow, on the way in, we missed it. We were confused and discussed that maybe we had forgotten the location, or maybe we had been watching something on the other side of the road. We were determined to stop during our departure. We didn't find what we were looking for. Instead, we found that road improvements had included two large culverts which had drained the small marsh. To install the culverts, the trees were removed. What had been a small patch brimming with biodiversity was now part of the cow pasture.

Yet, wildlife hangs on. I continue to walk near Mario and Raquel's urban home and continue to be surprised by what I find. A Gray Hawk nested in the same small ravine where I saw the motmot. A couple of visits back, I found the rare Cabanis's Ground-sparrow in a tiny patch of undeveloped land near my hosts' home. Alas, last visit, at least half of that habitat had been mowed and burned. Why be so neat? Wildlife prefer it wild. That's why I am going to Rara Avis. What is the secret to Rara Avis? Bad Roads. Very Bad Roads.

CHAPTER **2**

WHY AM I HERE?

> *Wilderness itself is the basis of all our civilization. I wonder if we have enough reverence for life to concede to wilderness the right to live on?*
>
> — *Mardy Murie,*
> 1998 Recipient of the
> Presidential Medal of Freedom

I wake up early and have breakfast with Mario and Raquel—always my favorite: *gallo pinto* (rice and beans Costa Rican style) and fresh fruit. I love papaya—difficult to have good papaya in Colorado. But the rest of my trip is going to be different. I have a comfortable home and life. I'm about to expose myself to many discomforts and a few risks. My reasons are both concrete and abstract. I want to see some antpittas. These are species of the dark parts of the deep forest. I want to learn more about them. I want to see them on my own. Part of my desire, perhaps, is ego. Seeing these without help of a guide or a feeding station would be a satisfying proof of my ability as a birder. More abstractly, these species are fascinating. There is nothing like them in the temperate forests where I grew up or anywhere in the United States. Only one species, the Scaled Antpitta, lives as far north as Southern Mexico.

I am also here because I want to feel more at home in the tropical forest. The only way to feel at home is to have more experiences. More experiences require more hours. I am going to a location where I will probably not see another person except at breakfast and supper. The rest of the time, I will be in the jungle, walking, sitting, and listening. I love to watch wildlife. I love to understand and marvel at the intricacies of evolution and the web-of-life. All this might sound trite, but that doesn't make it any less the truth.

My solo experiences in nature always lead to contemplation of the existential question, *Why am I here?* Sometimes I feel selfish. By being here, I've given up time with my wife, children and grandchildren. I often remark, upon returning home, that my improved sense and knowledge of tropical birds and forests are both useless and uninteresting to most North Americans.

So! "Why AM I here?" Am I simply here for personal pleasure? That thought carries guilt—consistent with my conservative religious upbringing. Fortunately, I've mostly moved beyond that, trusting my inner sense of what's right, and I am grateful for the good fortune of having a wife who recognizes the joy these experiences bring me. And, of course, she reaps some benefits from them when I return, renewed in spirit.

The contemplation of "why" soon becomes a flood of thought. I think of Darwin's famous statement that "there is grandeur in this view of life." I feel that grandeur. It has always been so. When I was a religious person in the conventional sense, it was always nature that made me feel closer to a deity. Indeed, I recall encountering the term "pantheism" in my youth and feeling that it best described my understanding of god.

Life is a marvel. My life, your life, everyone's life, are marvels. That's why I'm here. Much like Thoreau, who wanted to "get to the basics of life," that's what I'm doing. This is my meditation retreat. This is my religious retreat. This is what re-invigorates my love of life, the universe and everything. All this is why I'm here.

Although I have made more than forty trips to Latin America, I remain surprised by the simultaneous familiarity and separation I always feel. Here in mid-April, daily high temperatures in my hometown in Colorado are similar to Costa Rica's Central Valley. Both locations have mountains on the horizon. The difference is that Colorado's April seems poised on some fecund breakthrough, whereas Costa Rica's urban Central Valley seems to be holding it back. The houses, concrete, and people keep the jungle at bay. In Western Colorado, a post-apocalyptic world would revert to desert without the watering and care we give our plantings. Here, the jungle would quickly return.

The driver arrives on time. Danilo is familiar to me. He has worked for Rara Avis for many years, and I associate him with nearly every trip I've made. Because he's made the journey thousands of times, I'm probably just another rider, although I am a bit different, in that I am usually picked up at my friends' home and not at some hotel like most tourists.

We wind through the streets of Heredia in an easterly direction, catching a glimpse of Volcán Turrialba, which is belching a bit of smoke and ash this morning. Active volcanos are a part of life here.

One year we had planned to visit Volcán Poás but had to postpone because eruptions were too dangerous. Another time while visiting Volcán Arenal, we were treated to nighttime spectacles of automobile-sized glowing boulders blasted from the crater. These volcanoes are part of what is known as the "Ring of Fire" that arcs northward from New Zealand along the east coast of Asia, then easterly to the US Pacific NW and again southward, encompassing the west coast of North America and extending into South America.

What an interesting little country this is! As I noted in Chapter 1, Costa Rica, by some accounts, contains 5% of the world's diversity with 0.25% of the land mass. The great diversity is the result of The Great American Interchange--the joining of the Neotropics (roughly South America) and the Nearctic (roughly North America) ecozones to form the Americas. The interchange is apparent both from geologic stratigraphy and the dramatic effect on the many species (reptiles, amphibians, arthropods, weak-flying or flightless birds, and even freshwater fish) that migrated. The occurrence of the interchange was first discussed in 1876 by the "father of biogeography," Alfred Russel Wallace, who was essentially the co-discoverer of evolution with Charles Darwin.

Many evolutionary questions remain to be answered, but I stop thinking about them when, within minutes, we are on the main highway. Soon we arrive at the tunnel that changed Costa Rica's Caribbean Slope forever. There really wasn't any kind of road until 1960, but in 1983, with the completion of the Zurquí-Río Sucio Tunnel Project, travel to the Caribbean slope became almost trivial—except for the fog, and now the traffic.

The road cuts through Braulio Carrillo National Park. It is a complex maintenance nightmare because of the inhospitable location: abrupt topography, high mountains, dense forests, and heavy rain. Landslides or traffic accidents often close the road for hours or even days. Currently, this road is the only route to the Costa Rican Atlantic ports of Limón and Moín.

Now there is talk of building a four-lane highway to these ports. All of Costa Rica may soon be urban. I find myself wishing for the old, bad roads. The American nature writer Joseph Wood Krutch once noted that the natural marvels that still existed in his day in Baja California were a "a wonderful example of how much bad roads can do for a country." I was to remember that quote later this day.

The familiar highway descends toward the Caribbean Coastal Plain. We pass a parking area that accesses trailheads within Braulio Carrillo National Park, where I have hiked and looked for birds on several occasions. Then we pass the adjacent private reserve known as El Tapir—a great place to find the Snowcap. Now what would be a Snowcap in this tropical clime? It is a hummingbird. The male is reddish-purple with a snow-white cap. Resident from Honduras to Western Panama, it is unique in a family

(*Trochilidae*) replete with species almost too unusual to be believed.

Costa Rica has forty-eight possibly nesting, all reasonably-possible-to-see hummingbird species, as compared to the United States which has only four north of the shared border with Mexico. The Snowcap, as with all of Costa Rica's hummingbirds, has a shape North Americans would recognize. Even though some are called Coquettes (males have a small plume), Emeralds, and Mountain-gems, Costa Rica's hummingbirds appear familiar. But in South America, it gets weird. There are Trainbearers—little hummingbirds with five-inch tails, a swordbill—a tiny hummingbird with a four-inch bill, and a hummingbird called a giant—because it is as big (6 1/2 inches) as a swallow. These are just a sampling. Species range from the bright rust and orange Shining Sunbeam to hordes of drab brown or olive Hermits (Costa Rica has three Hermits.) that sneak into other bird's territories. There are Helmetcrests that thrive in the high, cold Andes above tree line and many diminutive Woodstars (Costa Rica has a single Woodstar) that act like bumblebees in order to delay being recognized and chased by more territorial hummingbirds.

El Tapir was once a butterfly farm, which accounts for the array of flowering plants. I've stopped there several times, and besides the Snowcap, I have visions of Green Thorntails and the exquisitely colored Crowned Woodnymph. The latter species, when seen in good light, is bright velvet-purple, except for an emerald green gorget that extends from under the chin to the belly. It is a stunner. But none of that for today. I'm headed for the deep and dark jungle.

I reflect on the fact that hummingbirds are only found in the Americas. America! That is a word I've learned to be careful with. I learned early on that Costa Ricans can be justifiably sensitive if those of us from the United States refer to ourselves as "Americans," with the assumption that the listener should understand where we are from. If we are to call them "Central Americans," that makes us "Norte Americanos." We have no right to be THE Americans. North America wasn't even the part of the "Americas" found by the first European explorers. Besides, some indigenous people don't much like the name "America," because that first mapmaker, Amerigo Vespucci, is associated with the Conquest and massive death and dying that accompanied colonization. Eighty to ninety percent of the indigenous peoples died, mostly from disease. Local extinctions were common (Chapter 27, *Lost City of the Monkey God*, by Douglas Preston).

Indeed, that death and dying may be the principal reason for the continued existence of much of the flora and fauna that I enjoy seeing in Central and South America. The late Jaguar researcher Alan Rabinowicz noted in his book, *The Indomitable Beast*, that this is the principal reason the jaguar, in contrast to other big cats around the world, still has a viable population in some areas.

So, what does this have to do with hummingbirds being found only in the Americas? Only that a proposal was made—and endorsed by some politicians in Ecuador, to eliminate the word "America" and replace it with "Colibria," because *Colibri* is a genus of widely-distributed hummingbirds. Why not? Hummingbirds are unique to what is now known as the

"Americas." Hummingbirds are universally enjoyed. They pollinate flowers and eat small insects. They thrive in almost every environment. Many are beautiful. "Colibria" as a name would be a uniter, never a divider.

Such are the thoughts that flash through my mind as Danilo and I pass the aptly named Rio Sucio (which means dirty river). The river's name refers to the sulfur deposits found on Irazú Volcano. The color of the water is a nasty, unappetizing yellow-brown. Irazú is the highest and most easily visited volcano in Costa Rica. On a rare, clear day it is possible to see both the Atlantic and the Pacific Oceans from Irazú. It is also still active—having famously erupted March 19, 1963 on the day US President John F. Kennedy began a state visit to Costa Rica. San José, the capital of Costa Rica, and much of the area surrounding the volcano were showered with ash, as has been happening recently, only from Volcán Turrialba.

The highway we are on, #32, soon reaches a major crossroads with Highway #4. This is terminology unfamiliar to many Ticos. They know everything by landmarks. My friends give addresses as something like, "a hundred meters north of the church and forty meters east of where the furniture store used to be." They find the idea of street signs and numbers ridiculous. I guess they haven't yet observed the Google mapping vehicle on their streets. Somehow, I can't see entering "a hundred meters north of the church" into mapping software. Ticos will conform as the rest of us have.

The crossroads at highways 32 and 4 is very familiar, as it is the way North to what my friends refer to as "the Sarapiquí."

The Sarapiquí River is famous for many reasons. It played a key role in transporting men and supplies in the battle against William Walker. The river also has hydroelectric dams, and along its banks are coffee and banana plantations. Those aren't reasons I'm interested, however. I'm drawn by its location in lowland tropical rainforest. The Río Sarapiquí flows through the famous La Selva Biological Station. With nearly 4,000 acres protected, La Selva has been in operation since the 1950s and has long been one of the premier locations for tropical research. Sadly, while La Selva still thrives, surrounding land use has affected the station's biodiversity. It simply isn't big enough, and that's unfortunate. Once, the research documented the inner workings of the tropical rainforest; now, some of it is documenting how species diversity declines as pristine forest patches become smaller and smaller.

Nonetheless, it is a critical piece in Costa Rica's biodiversity because it is part of a continuous slice of forest that extends to the mountains and the further reaches of Braulio Carrillo National Park. That's where I'm headed. Rara Avis, although privately owned, is contained in what is sort of a "bite" out of Braulio Carrillo's boundary. There is little access to this part of the park. The area is unusually pristine—apparently never having been logged. Rara Avis is, as some of their advertising says, "the real thing" when it comes to experiencing the rainforest.

We pull into the small town of Los Horquetas, where Rara Avis has a tiny office. I try on several pairs of rubber boots—my footwear for my time at the reserve—while the four-wheel drive tractor is readied and loaded for the ten-mile trip. This is my seventh visit to Rara Avis. Two years ago, my wife accompanied me on our sixth. Some love the trip, some hate it. Mostly, I tolerate and give thanks for it because I am thinking of Krutch's comment about bad roads. To some extent, I rue my selfishness. Because of the bad road, the reserve has never done well economically. Transportation costs too much, takes too long, and is too uncomfortable. But that's also why the Rara Avis rainforest remains so well-preserved.

The four-wheel drive tractor is hitched to a trailer that has two rows of seats welded back-to-back in the center. There is a bar across the sides so passengers can keep their hands inside. Then it is simply a matter of holding on.

The tractor is driven by Eduardo, a Nicaraguan. He says he has been doing the driving for fourteen years. I wonder what aches and pains the constant pounding has given him. The only riders today are me and a couple of people working at other locations along the access road. We drive rapidly through town. I smile and wave at children walking to school. They ignore me. Just another oddball gringo!

As we leave town, we must cross the Río Sardinal. I am flooded with memories. On our first trip, more than twenty years ago in 1991, there wasn't a bridge. We forded the river. I watched a couple of the local señoras washing clothing in pools among the boulders. That was a scene I never saw again—not in the modern Costa Rica. Eventually there was a small bridge, but it wasn't considered safe for passengers. We had to walk across. Not anymore. I also recall the wagon in which I'm riding being full of visitors. Now it is only me. I miss my family members. Rara Avis is one of the favorite memories for Mary and I, as well as for our son and daughter. My daughter and her husband even spent part of their honeymoon there.

I recall my son and daughter playing dominoes with the workers, helping researchers find and capture snakes, and squealing and diving in the plunge pools below the waterfalls. Now, they are busy with their families and careers. Mary says, "I can't do any wild wagon rides." Her chest is wired together following removal of a tumor from her heart, and the bouncing and banging are too uncomfortable. So it is just me. There is some melancholy with that thought, but I also feel freedom. I've always wanted to experience the jungle on my terms, as much as possible. Mary encouraged this trip. For her sake, I'm going to enjoy the heck out of it.

The local pair on the way to their jobs are Alejandro, young and good looking, and Vanessa, overweight and loud. He laid down on her lap, and they did a lot of joking and laughing, and what appeared to be some sexual touching, although Vanessa seemed older and twice Alejandro's size. I'm usually not distracted from the scenery and possible birds, but watching that pair was compelling. After initial introductions, they were oblivious to my presence.

At first the road isn't so bad. It has even been improved, I think. But then the road I recall emerges. It is difficult to describe. I have some very well-traveled friends who have been in backcountry all over the world. We have bad roads in Colorado—roads that attract off-road enthusiasts who enjoy matching their machines against the terrain. Nonetheless, both my well-traveled friends and I agree, the road to Rara

Avis is the worst road we have ever been on.

The road becomes mostly a route where no vegetation is growing. That's how you know where it is. That, and the fact, that years of travel have in some places gouged it deeper and deeper, such that the roadside rises several feet on each side of the tires. At times, one's view from the wagon is a red mudbank. When not confined by the mudbank, the road becomes a boulder field punctuated with holes so deep I once saw a worker step in up to his neck.

Frequently, our driver stops to do some roadwork with a pick and shovel. Many of the boulders show the shine of tire rubber from past slipping and sliding. In some low spots, the road has gotten wide, as the tractor has had to migrate farther and farther around a deep and growing pothole.

The Bad Road—Rara Avis Buildings in the Background

Sometimes, the road simply isn't passable. Workers from Rara Avis then must hike down and carry the tourists' luggage and supplies several miles to the lodge. The tourists walk. This leads to a story I find amusing.

It was 2006, we met two couples at the Los Horquetas Office one morning as we prepared to leave. The wives were very excited. This was both couples' first trip to the tropics. They had an arrangement: the men would plan part of the trip, and the wives were free to pick the rest. The men had picked fancy resorts in the La Fortuna area with its spas, hot springs, and nice restaurants. The women had picked Rara Avis because they wanted to experience "the real thing." We encouraged them. Because we had been to Rara Avis several times, they asked us a lot of questions. The women were genuinely interested in the flora and fauna, but I noticed the men asked about creature comforts.

Our experience had been that the upper floor of the "hotel" was the most convenient and probably most comfortable place to stay. We had rejected the "cabina" which was some hundred meters and a dark, often slippery walk from the hotel and open-air restaurant. The two couples were staying in the cabina. Mary and I wondered about their choice.

The next morning, at breakfast, things weren't well. The couples had encountered a tarantula, some amphibians, and numerous insects inside the cabina. They didn't like the walk. At least that was true for the men. The women still seemed game. They asked me more questions about flora and fauna. We told them about the more than one hundred species of orchids and the rare Stained-glass Palm. They talked with the resident guide about where they could hike. The men were off muttering.

At lunch, the women asked more questions, and we took them to see a few sights around the lodge and explained some things about the birds we were seeing. The men continued muttering. By nightfall, it was over. The men wanted out. *Now!* They complained bitterly to the guide and to their wives. A tractor must be sent. "We are leaving," they said, amidst discussion about hot springs, spas, and golf. What I haven't mentioned is that it had been raining. And raining! Over the radio, it was said that a tractor was sent.

I doubted the road was passable with so much rain. The guide agreed with me. "It is not coming," he said to me very quietly. I can still see the scene, as the clock ticked on into the rainy night. The two men were playing a card game to distract themselves, while their wives hovered nearby trying to avoid the icy stares occasionally sent their way. The rain continued to bear down. The couples had packed their gear and carried it from the cabina. It began to look as if they were going to have to carry it back—in the rain—and sleep there another night. The wives gently suggested they should give up and go back to the cabina. Nothing doing. My last recollection is when one of the men sensed defeat and began to agree with the wives. The other shut him off with a wave of his hand overhead. "Deal," he said. We went to bed.

I never did learn exactly what transpired. I went for a dawn walk, and when my family and I arrived for breakfast at 7 AM, the couples were gone. They had just left on the trail. A tractor had made it most of the way, and they were walking down to it and their escape. Not everyone liked Rara Avis as we did.

Now I am finding, as usual, that the last half of road is very, very rough. We bounce and slide and crash and eventually arrive at El Plástico. The name derives from the location's former use as a penal work camp. There were no buildings then. The prisoners were housed under a canopy of plastic sheeting. Once part of Rara Avis, El Plástico is now Selvatica—also a rustic forest reserve. We stop and visit with the caretaker. Despite wonderful access to primary forest and secondary forest with its own interesting flora and fauna, I don't think he sees many visitors either. This is where I get off.

The tractor will make it to Rara Avis today, but the last part of the ride isn't worth it to me. I'm eager to hike in the forest. The road here is nearly bottomless. There is a center ridge that the tractor and wagon ride on and slide on. Much of the way, the tractor's tires rub against the deepening gorge. It isn't fun to ride. It is loud. I want to walk.

I take a few cautious steps. I remember a previous trip when an obese couple regaled us with unlikely stories about their physical prowess. We started on the trail with them. The most overweight of the two, the husband or boyfriend, took about five steps before falling face first full-on into the mud. They didn't stay long either.

But, I'm happy. I like the feel of things. It is wet. It is warm. I like the fecundity of the jungle. Others have commented about how it seems to be "out there," eating and growing and ready to take over if unfettered. I like that. Now I'm listening.

What birds will I see? What animals? What plants? This is an adventure. I have the nine days to see what I can find, and I can't wait.

I'm immediately rewarded with a nice view of a Plain-brown Woodcreeper. Then some Carmiol's Tanagers flash by, accompanied by a White-shouldered Tanager. Then I spot a small flycatcher I'm unable to identify. When I complete my walk to the lodge, I find it is almost deserted. There is a woman—the cook. There is Giancarlo, the guide and chief. I say to myself, *Is that all?* In the past, there were usually other workers around. Not now. I find I can have any of the eight rooms in the hotel. I like the front corner upstairs room. I have stayed in it several times. This time, as shall be explained, it was a bad choice.

I refer to this building as the "hotel," but I don't want readers to have the wrong idea. The building is an eight-room box with a room at each corner on each of the two floors. There is a small spiral staircase in the middle leading to an open area with a nice view of the jungle on the second floor. The building was constructed of native wood harvested onsite. It was built by hand.

Beds are simply wooden frames upon which are placed small mattresses. There is a small deck containing a hammock outside each room. All that separates the room from the outside are screens—now with holes in them. Some holes are quite large. There is a familiar-looking bathroom with a toilet and bathtub/shower. Hot water allegedly has been available, but never while I visited—or maybe I had become accustomed to the cold water and just didn't bother checking. The water supply is the nearby river, the water merely piped into the building by means of gravity from the hill behind.

The running water is the only modern amenity. There is no electricity. In early years, we were given a kerosene lantern. That's why this hotel is version 2. Version 1 was burned down by careless handling of a lantern. I have good flashlights. There is a generator which provides lights for the open-air building that is used for cooking and eating. That building is some thirty meters down the hill from my room. The generator operates approximately four hours per day and only in the evening. I will bring my electronic devices down there every evening for charging.

I'm not certain what to expect regarding meals, so I have also brought quite a bit of food. For one thing, I intend to be in the jungle all day, not returning for lunch. In the early days, my family enjoyed the meals immensely. One of the early cooks even left to start his own successful restaurant—a matter of pride for Rara Avis because of the desire to be both an employer and a training facility for the local community. Later, the food, although still palatable, became less varied. The Costa Rican tradition of *gallo pinto* for every breakfast is to be expected. No problem there, but dinner became a mystery meat patty one night, followed by some sort of stuffed noodle the second—then repeat. I never spent a sick day at Rara Avis, but my wife has on a couple of occasions. One of our friends also related becoming so ill she had to be taken out as an emergency. Fortunately, I don't worry much about getting sick and will stick mostly to *gallo pinto*, fruits, and vegetables.

Giancarlo tells me there are no guests scheduled during most of my stay. Two guests are here now, a young couple: Georgina from Barcelona and Benjamin from Germany. I am the only one with English as a first language. There are some researchers from Florida State staying in some of the workers' cabins, but they went to the city for a few day's rest. Giancarlo knows I'm interested in birds and tells me the researchers had seen a Red-throated Caracara the previous week. This is a very rare bird in Costa Rica. I'm really interested.

Having a guide at Rara Avis has always been a hit-or-miss proposition. The first time we visited, in 1991, the late Amos Bien, the owner/founder, had come up for a few days. He had trouble finding workers because it was Semana Santa (Holy Week), when much of Costa Rica simply shuts down. At other times, there was a permanent guide, but it seemed as time went on there was less emphasis on guiding.

After a several-year hiatus, we returned in 2006. Birding friends of ours had told us the guide, Wilbur, was extraordinary. He was—in many ways. He was self-taught. He had an inadequate pair of binoculars. But he knew the plants, animals, and birds remarkably well. Many guides today use playback to attract or keep a bird in view. I'm sure that sort of equipment was outside of Wilbur's budget. He simply mimicked the calls—precisely. I learned that Wilbur had been hired as a simple worker but used his free time to study the lodge's reference books and learn everything he could about the flora and fauna and the English language. Once, I came upon him studying a dictionary. As with several other Rara Avis graduates, Wilbur eventually moved on to a long-term guiding career at one of the traditional lodges with more access to civilization.

It is sad to me that there are so few guests. Over the years, we have met many interesting people. I can recall guests from all over the United States, Canada, Ireland, the United Kingdom, Germany, and Scandinavia. On this trip, besides the couple from my first night, eventually there were two North Americans, two from the Netherlands, and two Swiss.

Ordinarily, I would spend more time with the paying guests, but once the Florida State researchers return, I quickly attach myself to them. The four are Megan, who is performing her PhD research, and her three field technicians. The young lady's name is Kim, there is Jason, who has the usual, scruffy look of a graduate student, and a young Italian named Luca. He lets me know that it is very unusual for an Italian to be a biologist. "They don't care about nature in my country," he says. After making a trip there a couple of years later, I had to agree.

Having the researchers here is amazing luck. Their day starts in the pre-dawn, so they can reach their blinds to watch displaying White-ruffed Manakins. To accommodate their early start, the cook arises in time to prepare breakfast at 5 o'clock. It is not an issue for me to join them, thus solving a major problem that concerned me. I had already planned not to return for lunch, because I want long sojourns to the remote areas of the reserve. Typically, breakfast is at seven, because that's what most guests desire. That would have meant I couldn't start any of my hikes until nearly eight, or else I would have to miss breakfast and eat the same crackers and tuna I am having for lunch.

The researchers are a pleasant group and tolerate my vicarious desire to be a tropical scientist. I grew up in a very narrow world because of my family's economic and educational circumstances. Fortunately, my parents had provided every opportunity they were aware of for me to broaden my proverbial horizons. Nonetheless, I saw my future as either needing something to sell or working in some sort of factory. I chose chemistry for my college major, in part because there were many chemical plants and refineries near my hometown. Obviously, these would be places I could work.

In college, I realized I might need some engineering classes if I was to work in one of the chemical plants. The one semester I had appropriate classes I ended up dropping as many as I could, while still remaining a full-time student (necessary to keep from being drafted and sent to Vietnam). I was relieved to earn C's in the rest—the only C's I ever received in college.

Eventually, I spun my major away from engineering and then from pure chemistry to pollution monitoring and geochemistry. That combination opened opportunities for fieldwork and working with field data. I had a successful career, but as soon as I became acquainted with the tropics as an adult, I wondered what would have happened had I visited a tropical forest with my parents or had done a semester or summer course in the tropics while in college.

Anyway, for now, I am accepted into the research group. I am thrilled! We take our meals together. I understand most aspects of their research and believe I am joining their conversations without embarrassing myself. At least they are nice enough to never let on if I say something stupid. It is also delightful because any questions I might have about my own day's observations can usually be answered by one of the four biologists. Besides, how great is it when the topic of the day stems from one of them returning with the news, "Today, I saw a copulation!" I never have understood why anyone would major in business!

Biologists often visit Rara Avis. Another time, Mary and I met a guest who was updating a guidebook about Costa Rica. He was very impressed with Rara Avis. He considered the surrounding jungle the most unspoiled and accessible in the country. Indeed, it is. More than 95% of Rara Avis is primary old-growth rainforest, located in the center of a small section of forest that apparently survived intact during the Pleistocene glaciations of the last 2.5 million years. This biologist was a birder as well, so we were both checking out various locations for rare birds.

One we had been interested in was a Purplish-backed Quail-dove. Quail-doves have been compared to vampires for their love of dark places. I have organized several group trips to the tropics. I always put together information on birds we may see. With respect to Quail-doves, I wrote, "You will not see a Quail-dove." But on that occasion, a pair had been known to check out the lodge's dump. The dump was in a steep ravine about 20 meters behind the restaurant. There was a small building that housed a generator that served as sort of a blind, so it was easy to sneak back there and check it out. Other birds and some mammals such as coatimundis also liked the dump, so whenever we were around the restaurant, one of us would go back and peek.

At one point, the biologist/writer came running back to get me. "There's a tapir back there," he said. Now the jungle encompassing Rara Avis has a lot of tapirs, judging by their frequent tracks. On the other hand, it didn't have any judging by my sight records. Indeed, I knew there had been few sightings. Tapirs are very shy, and despite their ungainly appearance, are quite fast. Shaking bushes and disappearing hind ends accounted for virtually all sightings. We were excited. Mary and I sneaked back and hid behind the generator shed. There it was—an apparently sleeping tapir down at the bottom of the gully. We watched and waited. It didn't move. Soon the biologist joined us. He said an actual photo of a tapir would make an outstanding cover for his book. He would wait until the tapir awoke, no matter how much time was required.

Mary and I backed away, leaving the writer to his vigil. We discussed the tapir and decided not to go back for fear we would scare it away. Hours passed. No biologist/writer. Finally, I returned to find him still at his vigil. He shook his head. "Maybe it is dead," he whispered. I waited awhile, and then left, not being interested in a wake for a dead tapir. Not too much later, along came the biologist, dejected, his great opportunity missed.

As we stood there talking, here came the tapir! It walked into the clearing next to the restaurant. One of the workers bolted from a nearby shed. "Miguel," he shouted to the tapir. "Miguel!" The tapir turned and walked over to him and lowered its head as if to be scratched—which is what it wanted—and what the worker did. The worker ran off and found certain leaves and fed them to the tapir, which soon stretched out on the ground and let all of us scratch him as he snoozed.

Now the story came out. The tapir, as a juvenile, had been attacked and badly wounded by a jaguar. It had crawled into a shed at Rara Avis. This worker had found it, nursed it back to health, and named it Miguel. What was fascinating is that some researchers had put a radio-collar on it for a while. They found that Miguel was as wild as any other tapir out in the forest, but if he happened to pass near the clearing, he walked in and became a pet for a few hours. That's how I came to have my photo taken scratching the head of a tapir, sleeping, with its mouth open and its tongue lolling on the ground.

This story has a sad although possibly appropriate ending. I returned a few years later and asked if Miguel was still around. They pointed to the "rancho," a small building used for meetings. Up on the wall was a skeleton. Miguel had been killed and eaten by a jaguar. They had cleaned his skeleton and mounted it on the wall.

Miguel, the Friendly Tapir

Truly wild and abundant tapirs are a fraction of the fauna and flora in this jungle. There are more than five hundred species of trees, one hundred species of orchids, and other rare and interesting plants such as the "stained-glass window palm (*Geonoma epetiolata*). Rara Avis has a bird list of 367 species and a list of mammals exceeding forty, including (more about this one later), the Watson's Climbing Rat.

Moreover, Rara Avis is an important altitudinal corridor for the movements of endangered jaguars, pumas, quetzals, and Great Green Macaws. It protects the borders of Braulio Carrillo National Park. By protecting its forest, the Rara Avis reserve, according to its brochure, has avoided the emission of over 337,000 tons of carbon dioxide that would have been emitted had the forest been cut, as was planned by the previous owners.

I'm here! I'm back. I'm so pleased to be here I can hardly stand it. Tomorrow, I get to spend all day in the jungle! I can't wait.

CHAPTER **3**

IMMERSION

There's a common misunderstanding among all the human beings who have ever been born on the earth that the best way to live is to try to avoid pain and just try to get comfortable. You can see this in insects and animals and birds. All of us are the same. A much more interesting, kind, adventurous, and joyful approach to life is to begin to develop our curiosity...
--*Pema Chodron,* Awakening Loving-Kindness

To be worthy of a happy immortal life, we must in this present life be careful to bring no avoidable pain or hardship to any creature, and so far as we reasonably can, to ease the burdens and promote the welfare of all that surround us.
— *Alexander F. Skutch,* Quest for the Divine

Exploring the jungle is about curiosity. I suppose exploring any environment, any person, or any subject is about curiosity. I always look to a quote from Eleanor Roosevelt, "I think, at a child's birth, if a mother could ask a fairy godmother to endow it with the most useful gift, that gift would be curiosity." With that in mind, I have wondered if my curiosity about the jungle is intellectual laziness. The tropical naturalist Marston Bates, when contemplating the natural mysteries of the jungle, noted that "there is as much in a square foot of his backyard." Yet, the backyard curiosities are tiny and familiar. One cannot avoid a sense of wonder about the jungle because no matter how passive you are, an insect will land on you or bite you or an unusual plant will catch your attention.

Seeing antpittas was my tangible goal for being here, but I was already sidetracked by the Florida State Research team's sighting of a Red-throated Caracara. Giancarlo also wanted to see it and offered to accompany me for the day. Thus, finding that rarity became my first objective.

Although it happens every time, I was immediately frustrated because I was enveloped in unfamiliar sounds. I compare the feeling to being in a foreign airport where the signs are unrecognizable and the announcements over the Public Address system unintelligible. Other people are reacting to the sounds, but you have no idea what is being said. There is a feeling of desperation, knowing others have this information you desire. That's the way the jungle feels to me at first. Because I have a hearing impairment, learning bird calls has never been easy for me, but I will soon remember a few.

What I hear initially is a loud two note call—long on the front end, short on the back. This, I remember is the Bright-rumped Attila, a flycatcher that lives in the canopy. This call has an easy mnemonic: "eat-it, eat-it, eat-it, nowwww" it says incessantly. I soon pick up another: the Green Shrike-vireo. Its call of *"peeta-peeta-peeta"* is somewhat reminiscent of the Tufted Titmouse of the Eastern and Midwestern United States. Although I will hear both almost constantly during my stay, I shall see neither. The Shrike-vireo is particularly elusive and was a long-time puzzle for the scientist, author and naturalist Alexander F. Skutch. As he says, "I dwell here surrounded by mysteries...a clear whistle, rapidly twice or thrice repeated, is borne to my ears from the neighboring forest. I know it to be the song of the Green Shrike-vireo; for the next few months, I shall hear these notes innumerable times each day; I shall scan the lofty treetops until my neck aches for a sight of the elusive bird...most probably in vain. Year after year, I hear these same wild wood notes yet for more than three decades, I have searched fruitlessly for this bird's nest, which to my knowledge has never been described."

Skutch is a hero of mine. I have read his many books, and they have inspired much of my love of the American tropics. Skutch died in 2004, a week before his 100th birthday, having lived in Costa Rica's Valle de General (near San Isidro) for much of his life. During his time there he authored 30 books. I have tracked them all down, and they watch over me as I type. Skutch was a polymath and an ascetic. I desire to be the first, probably not the second.

He was a man of almost infinite patience and equanimity. I love his books for a myriad of reasons. Initially, Skutch wrote "Life Histories" of tropical birds. He took advantage of a fact, always true in scientific research, that "being there first," is the best way to be published. That is, if you describe a new phenomenon or visit an area first, you have no competition for your findings. Publications and fame soon follow.

He began his career as a botanist, but eventually switched to ornithology when he became enamored of birds while watching the common Rufous-tailed Hummingbird build a nest. Skutch realized that most birds had been collected and named, but little was known of their nesting habits. With that realization, his life's work was established. He had little funding, financing most of the work with botanical collections. He traveled through much of Latin America with sojourns to Guatemala, Ecuador, Panama and Peru before settling along the Río Peñas Blancas in Costa Rica's Valle de General.

Skutch took few photographs, and he didn't collect (that is, shoot) birds. He performed his research by observation. With a crude blind and a notebook, he spent countless hours simply watching. In this case, "countless" isn't a throwaway term. His books are replete with details such as the average number of minutes males versus females sat on eggs, or the relative feeding tendency, including identification of the food. From his meticulous observations, much was learned regarding nesting behavior, answering such questions as relative male versus female participation, days until hatching, days until fledging and so on. He also gathered a lot of data on nest failure and was quite saddened by the knowledge that most nests (perhaps up to 85% in some species) failed. Predation was a severe problem, especially by snakes—the one group of native inhabitants towards whom Skutch harbored some enmity. Indeed, he dispatched marauding snakes on several occasions. True to his personality, Skutch took some solace in the low rate of reproduction by realizing that the adult birds themselves must be relatively long-lived or the species would soon disappear.

Some of his work has been disdained by more conventional ornithologists. As I noted, Skutch would not collect birds, and, at times, he anthropomorphized their behavior. In my view, he did this out of the deep appreciation he had for most of nature, and his belief that evolution tended toward harmony rather than conflict. For that reason, Skutch saw predation as a miscarriage of evolution.

Indeed, late in life, he wrote a commentary suggesting that "Biodiversity has certainly become excessive and is responsible for a major part of the sufferings of animals, including humans." His idea for conservation of species in a shrinking world was to eliminate or carefully regulate predators and to promote "biocompatibility instead of unlimited biodiversity." He hoped such a natural world would preserve "a friendlier, more peaceful ambience." Considering that the most lethal bird predators in the US are feral cats which are, apparently, defended by millions, his ideas, while noble, are hopelessly naïve.

I found it ironic that Skutch never saw the point of view of the insects the birds eat. He was aware that the beautiful Blue Morpho Butterflies were prey of favorite birds of his such as jacamars and puffbirds. Birds do have relationships with fruits that are often symbiotic—but even there, exceptions occur. Flowerpiercers and some of his beloved hummingbirds have learned to pierce the sides of flowers and steal nectar without contacting the pollen such that there is no benefit to the plant. So much for harmony!

Still, Skutch was not one to give up on humanity, something I often feel inclined to do. After reading his works, I can see that his desire for harmony in the world was paramount. He looked at humans and their destructive impulses and was filled with consternation. If we could just sit back and be fascinated by nature's beauty, wouldn't that be enough? His idea was to cultivate gardens and only include inhabitants that could live together without discord, hoping that we humans might be inspired to do the same.

In any case, Skutch was the first to describe nesting behavior for many species. I enjoy reading these accounts, not because I am so interested in how many minutes a female Barred Antshrike brooded the young and whether or not the male or female sat on the nest overnight; I enjoy reading them for the small grains of description of how the environment was in Costa Rica in the 1930s and subsequent decades before being overwhelmed by development. I love such accounts because I don't want humans to forget what this planet was like. When I am enjoying what I see now, I want to understand how it once was. "Nature is like a book," I once read somewhere, "except the finest and best pages have already been torn out and discarded." Skutch's writings provide access to those "finest and best pages."

Looking Out Skutch's Window

The other gems I enjoy finding in Skutch's writing are related to the equanimity he possessed as he documented how so much that he loved was either killed or damaged by man or as part of nature's usual cycle. As someone who has often let a single negative condition color everything in my life, I was encouraged that Skutch could remain at peace even as he watched the destruction of that which he most cherished.

Here are a couple of examples. Dogs barking and running in his forest meant that animals and birds were being unnecessarily harmed. This hurt. Here was man, being careless with his domesticated animals. Here were the dogs, at loose in an unfamiliar environment, free to kill and terrorize anything they found. The dogs had no part in any sort of "balance" of nature. Skutch lamented the destruction but he never gave up. Instead, he said, "If to expect to realize our dreams in all details is a wildly extravagant hope, to make no effort to fulfill them is to relinquish precious opportunities for enriching our lives…I shall say little more of the tribulations that befall one who tries to preserve nature amid neighbors intent upon plundering it…"

On another occasion, Skutch severely injured a shoulder. It was many weeks before life was more than a painful ordeal. Yet, he noted, "[I]n order to experience the sensations of returning health...one might willingly suffer a fair amount of pain…[C]onvalescence has much in common with adolescence: both are periods when we feel our powers daily waxing. Convalescence carries us back to that time of youth when life was full of promise." Truly, Skutch's hold on equanimity was incredibly strong, something I've always tried to emulate. That probably surprises my family. I have tried, but equanimity does not come easily to me.

And, it wasn't as if Skutch was Pollyanna. As he endured his pain, he found the promise of new discoveries in ornithology "generously fulfilled." Yet, adding to the pain of his shoulder was "returning to rarely found nests only to learn that some predator had pillaged them." At the same time, he experienced the "deep distress of watching my landlord's brother, a lad of twelve, waste slowly away and die horribly inch by inch always in agony, from an infection…despite all that could be done locally to relieve his pain." Skutch wisely noted that, "The blue threads of happiness and the red strands of pain are so tightly interwoven in life's fabric that…not even the wisest and the best of us succeed in disentangling them."

I also appreciated Skutch because of his relentless pursuit of the meaning of life. Most know him today as co-author of the first comprehensive bird book for Costa Rica: *Birds of Costa Rica,* co-written with Gary Styles. This book, published in 1989, while not having the best illustrations for the birds, will always be a useful reference, because it is rich with insights into behavior—something impossible for smaller guidebooks to provide. But Skutch's writing and seeking went well beyond birds. I've always believed he should be celebrated for his books on religion and philosophy: *The Quest of the Divine, The Golden Core of Religion* and *Life Ascending.* In these three books, Skutch relates his quest to learn which religion had the best answer for the meaning of life.

All three of these books are a tough read. They are deep and lengthy. I couldn't read any of them for long before my chin dropped to my chest. The prose and subjects are arcane. I read them over many months, taking notes as I went along. Skutch pored over texts and interpretations from every religion I had heard of, and some I had not. Christianity was only part of his study as he plunged into Hinduism, Islam, and Jainism. He never found what he was looking for, that is, a comprehensive system of belief and action that was consistent with the natural world he observed and the peace and harmony he believed was the destiny of humankind.

I do not wish to offend anyone in my agreement with an observation from the character Gus in the hilarious, yet poignant, coming-of-age novel, *The River Why* by David James Duncan. Gus, when referring to the Bible, says, "I have learned that no book ever published has generated so much tension, division, and strife...as that conglomeration of myths, laws, genealogies, songs, visions, poems, histories, letters and tall tales known as the Holy Bible." This quote made sense to me, but it always felt flippant and like a cop-out until I read Skutch's painstaking attempt to find the essence of the Bible and other religious texts. He concluded that, "{religion's} tragedy is that, in the absence of adequate knowledge, it has been compelled to make guesses and assumptions about the things which transcend mankind, and all these assumptions...have crystallized into hard dogmas; and the irreconcilability of these dogmas has stirred up fierce conflicts among men who should be united in a common endeavor." Yes, exactly that, in my mind!

I was raised in a conservative Roman Catholic home. Our parish had a conservative monsignor and conservative nuns. Certainly, they meant well, but all I recall is being constantly reminded that our community did not produce the number of priests and nuns commensurate with our population. "Some of you are refusing your vocation," we heard. I was sensitive. I wanted to be good. I went to church a lot—even on Saturday. I "did" the "Nine First Fridays" and "Five First Saturdays," or whatever they were. My mother let on that she was "praying for my vocation," and let me know years later that she actually thought I was going to be a priest.

I tried to believe and conform, I really did. When I left Catholicism, I embraced a liberal form of Protestantism. I taught Sunday School and led youth groups. I also studied. As Hamlet said, "...ay, there's the rub." My career was in scientific research. To be a better scientist, I read all I could about my fields of study. To be a better Christian, I tried to do the same. I probably read some dozen books on Christian and

Bible history. Held up against my scientific training, that history didn't jive with the current practice of Christianity.

Seminal to my rejection of Christian dogma was my teaching a class on the Book of Mark. The course of study I was using simply dismissed portions of the book, saying, "Scholars believe those verses were added by the early church to punish or frighten a rival sect with which they were in conflict." I didn't take that statement on faith; I had to learn why. I learned that the apparent "eye-witness" accounts were written generations after the events occurred. Sections of the Bible were added and subtracted at a conference where most of the invitees didn't come, and as one scholar put it, "There was wine!"

Eventually I read books such as *The Power of Myth,* by Joseph Campbell, as well as most of his other works. It became obvious. The Bible is a collection of good stories, the roots of which predated the time of Christ. Then there was the obvious matching of pagan with Christian holidays. And, finally, there are the historians who have shown that the form of Christianity that finally won out in the old days prevailed because its adherents wrested economic power from other Christian sects.

Like Skutch, I realized I did not have any religious faith. As soon as I wrote that word, I opened my browser and typed it in, and this is what came back: "Faith: strong belief in God or in the doctrines of a religion, based on spiritual apprehension rather than proof." Skutch wrote about what he observed, not what he wanted to be true. Likewise, in my professional life I conducted research. Publishing in scientific journals was a requirement. Publishing meant passing peer review, which meant I had to "prove" what I was saying with data.

Like Skutch, and later like Daniel Dennett in his exquisitely presented *Breaking the Spell*, I asked how could it make any conceivable sense to suspend reliance on facts for something deemed to be as important as religion?

And yet, like Skutch, I am a seeker of a philosophy of life consistent with my understanding of the human condition. That's why his books on religion so impressed me. That is why I can't believe in some higher power without evidence. Indeed, the evidence points in the other direction. If my belief system can be characterized, it is vaguely Buddhist, but not as a religion, only as a philosophy of life.

Buddhism is replete with its Four Noble Truths, Eightfold Path and so on. I have whittled these down into something much simpler. The most important is that bad things will happen in everyone's life and most are not the person's fault. The second most important factor is that the best way to manage the bad things that happen is through kindness to yourself and fellow man. That's it. Isn't that enough?

Yeah, that's enough!

When I look at the direction this chapter has taken, I recognize that solo time in nature is a religious experience for me. And, as Skutch says, "the tropical rainforest is the highest expression of the creative movement on this planet." But then I realize the statement I just made is too narrow. If I understand Skutch, all of life is to be a religious experience—in the sense of being appreciative and contributing to harmony. Because that statement makes me feel like such a failure, I realize that's why I admire Skutch. He seems to have done the best job

I know of living his life that way. I think back to his "blue threads of happiness and red strands of pain." He couldn't untangle them. None of us can, but we must try, though we know we can't fully succeed. That's the meaning of life.

These thoughts are running through my mind as I walk with Giancarlo. Such thinking is typical of my initial immersion in the tropics. My mind races. I go "all in" on philosophy of life as I wrestle with the random thoughts. *It is selfish for me to be here! What good is doing this? I've left my wife alone. I'm not at home. I should be reading or playing with my grandkids.* The "shoulds" pour in—for a while. It is like an echo—loud at first, then soft, until all I can hear and sense is the jungle. That's better! Thoreau said, "What business have I in the woods, if I am thinking of something out of the woods?" Why can't I do that? It always is the same, but after a while, my mind becomes more focused.

We hike the trails where the Red-throated Caracara was sighted. These birds have noisy parrot-like calls. We hear nothing of the sort. I play the call. There is no answer. Birding seems difficult today. The jungle is thick, and the birds are high in the canopy. When we finally spy a mixed flock, I identify Scarlet, Summer, and Hepatic Tanagers—all common in certain locations in North America—although typically not possible to see on the same day. I am disappointed. I came here to see "tropical" birds. Then I remind myself, *I just did!*

This is April. Those tanagers are headed North. The Summer Tanagers I just saw are rather common in the Southern United States, but only from late spring to early fall. We in the United States have no right to refer to Summer Tanagers (or any other neotropical migrant) as "one of our birds" unless we are including the rest of the Americas.

If there is a concept I want to impress upon North Americans, it is that we are one world, and the birds show us the way. Let's consider a couple of other familiar birds. Almost everyone in the United States can encounter a Yellow Warbler. They breed in almost every state and are common migrants in the few Southern States where they don't reproduce. However, most live in the US only about four months—the rest of their lives are spent in Central America and Northern South America. Whose birds are they?

How about the Eastern Kingbird? I remember seeing them all summer when I lived in Southern Illinois. They would be on the fences and wires around the farms where I worked in the summer. The trees around the golf course I played always had a few. Try to find an Eastern Kingbird in the United States in April, or in September. It isn't impossible, but there won't be many, and you can forget about it in March or October. They live in the tropics for 8 to 8 ½ months per year. In the US, the Eastern Kingbird is known for its insect diet, usually sitting on a prominent perch and sallying out for a meal. They are also aggressive and territorial, and frequently seen chasing much larger birds away from their home area.

Another common migrant is the Swainson's Thrush. It's habits in the US could hardly be more different from an Eastern Kingbird. This thrush is a skulker. One guidebook refers to its "solitary" habits. It commonly nests on some mountain property my family owns in Colorado, but I usually work hard to see one. The Swainson's has a song often termed ethereal. It is flute-like with a spiraling quality—quite beautiful as it seems to emanate from dense vegetation. Unfortunately, the thrush only seems to sing for a week or two. It is shy and cryptically colored. The Cornell website "All About Birds," refers to it as "arboreal," and having a diet of mostly insects.

It is different in the tropics! Both the Eastern Kingbird and the Swainson's Thrush change their diet to include lots of fruit. When there is fruit in a tree, there is plenty of it, and mixed flocks can get together and eat until satiated. Hence, in the tropics, both birds are commonly seen in flocks. One April morning while hiking at the Organization for Tropical Studies Las Cruces field station in Southern Costa Rica, I had eight Swainson's Thrushes in one binocular view. They weren't arboreal either. They were all on the ground. In April, the Costa Rican jungle can be full of them. If they are not the most abundant bird, they are the most easily seen. Eastern Kingbirds also occur in large flocks eating fruit.

Completing the contrast, later this trip I spied a Swainson's Thrush, a Wood Thrush (a rapidly diminishing breeder in the Eastern United States), and a rare tropical Blue and Gold Tanager eating fruit from the same bush. Clearly, we in the US need to change our myopic perspective of not thinking about what "our" birds are doing when they are not here. Their lives are different, yet no less rich. They are dependent on a part of the world that we must pay more attention to—if we want to continue hearing that ethereal song of the Swainson's Thrush and continue to be impressed when a pair of Eastern Kingbirds dive-bomb a much-larger Red-Tailed Hawk.

But back to today's walk! Eventually, our wandering takes us on a trail along the stream next to the lodge. Giancarlo tells me a Lanceolated Monklet has been calling in the area. Now there's a bird I haven't thought of. Not much is known of it. Stiles and Skutch's *Birds of Costa Rica* lists the nest as "undescribed." Under "Habits," their book says, "little known." Under voice, "Usually silent." It is also mostly brown and only about five inches long. The Cornell Online Laboratory of Ornithology has a brief physical description and then under "Life History," every topic says, "No Information Available." Having read such descriptions or lack thereof, I had filed this bird in my mind under a category like the Quail-doves: "You won't see it."

The name Lanceolated Monklet must sound quite strange to those unfamiliar with tropical birds. Monklets belong to a family known as puffbirds, because their short tails and relatively large heads make them appear "puffy." Puffbirds are confined to the tropics and belong to the family *Bucconidae*. The Lanceolated Monklet is the smallest member of the family and the only member of its genus (*Micromonacha*). Apparently, puffbird diets are almost solely insects with occasional small amphibians. As a group, the thirty-some species are seldom seen because of their seemingly lethargic habits. Many tropical species that feed exclusively on insects are very active, always foraging, always checking bark and leaves for food. In contrast, puffbirds hunt by ambush. Their bodies may not be moving, but their eyes are watching. When an object of prey moves in their range of vision, they make a rapid foray to grab it. They return to their perch, consume the victim, and resume waiting and watching.

A few in this family, such as South America's Russet-necked Puffbird and the widely distributed White-necked Puffbird, will frequent open country and forest edge. I had some great encounters with the former in Colombia where a pairs' preferred perch was on fenceposts. Though they moved little, they were easy to see.

The Lanceolated Monklet, however, does not like open country, and it is rare. Its preferred habitat is dense vegetation along wet and dark, often-steep, pristine mountain ravines. Thus, their habitat is diminishing, and what remains is difficult to access and difficult for viewing. For these reasons, I presumed I'd never see one.

I have recordings of all of Costa Rica's birds. Although the monklet is usually quiet, I play the call. According to Stiles and Skutch, the song consists of "high, thin, plaintive whistled notes with rising inflection, each slightly higher pitched than the last." It is distinctive. I play it a couple of times and hear nothing but the stream roaring below.

Giancarlo and I walked and listened all day. No Red-throated Caracara. No Lanceolated Monklet. Now, my first jungle day is all but over. I return to my room and sit on the balcony. I enjoyed the company of Giancarlo, but I hadn't had the silent one-on-one reverie with the jungle I had anticipated. I felt good but the total immersion I craved hadn't quite happened. There is an hour or so until darkness. I'm not tired. Why not revisit the trail across the bridge and above the stream?

What I am thinking of is encountering a Black-faced Antthrush. Chicken-like in shape, but small, this antthrush lives near the ground and can be difficult to see except for its loud and distinctive call and a frequent habit of walking on open trails at dawn and dusk. Seeing one would be a nice re-acquaintance with a characteristic jungle species. And, I know, I want some time, in the jungle, alone.

I move slowly but hear and see nothing. With nothing to see or hear, I walk farther than intended, and realize I am in the location where that morning Giancarlo said a Lanceolated Monklet had been heard. I play the recording. I play it a couple of times expecting nothing. After a few moments, I look at my device. It isn't playing, but I am still hearing the Lanceolated Monklet! Now I have to find it. The worst part of my impaired hearing is not that I don't hear sounds so well, but that I have a difficult time with direction. Everything sounds as if it is in the direction of my good ear.

That's right. I'm a birder with bad hearing. Oh, my right ear is normal enough for an old guy, but the left—not so hot. I have some residual hearing, but almost nothing in high frequencies. Thus, as I try to triangulate the location of a sound, I'm apt to spin around, because the sound always sounds as if it is to my right. The problem is especially bad if the sound doesn't repeat or is overhead or if there is a lot of "white" noise, like the rushing stream at my present location.

That hearing loss has been important in my life. Once when hunting with my dad, he told me he'd fire two quick shots when I should walk over to meet him. I was a teenager without much understanding of my hearing handicap. Dad fired the shots and I headed the direction from which I heard them—my right—the side all sounds came from. I went the wrong way, and it was quite some time before we met up. I was angry with him for misleading me, before realizing that it was me who went the wrong way.

At one point, I was delighted by my hearing loss because I was of draftable age in the time of the Viet Nam War. I was sure my hearing would disqualify me. Not so. To my chagrin, it disqualified me from the Navy and the Air Force and from all officer training. The only thing I was qualified for was foot soldier in the Army—exactly the circumstances where my one-sided hearing would put me at most risk. Fortunately, I was able to escape service, but that's another story. Now I have to overcome my handicap and find the monklet.

The bird is agitated, and it keeps calling. I keep looking. This is my one and probably only chance to see this very rare bird. I can't find it. I keep sweeping my right ear from side-to-side, trying to locate the source of the sound. No luck. I begin to panic. How could I be this close? How could I be so incompetent such that I could hear it *right here* and not find it? The vegetation is very thick. The river is very loud. The darkness is gathering quickly. Suddenly to my relief, the bird is there, right in front of me on an exposed branch. I can watch its mandibles part as it trills over and over to its unseen rival.

It is oblivious to my presence. I have a good look at the well-developed white, facial bristles (as with many insect eaters, the bristles may protect the eyes or help hold struggling prey), its eye-ring and strongly streaked chest. It appears to be almost all head—that puffy look common to puffbirds. It is a stout, cute little guy, with a large black bill—perfect for snatching and devouring a large insect. Now I feel a little bad. The bird is putting on a great show. My playback of his call has upset him. Either he thought I was an intruder he had to vanquish, or maybe he thought I was the mate he was looking for. In any case, he keeps singing as darkness descends—which, in the jungle, near the equator, happens very fast. Reluctantly, I turn, then go back for one last look, and still hearing the plaintive whistle, return elated, and immersed, to my room.

CHAPTER **4**

PATHWAYS

> *I have often noticed that these things, which obsess me, neither bother nor impress other people even slightly. I am horribly apt to approach some innocent at a gathering, and like the ancient mariner, fix him with a wild, glitt'ring eye and say, "Do you know that in the head of the caterpillar of the ordinary goat moth there are two hundred twenty-eight separate muscles?" The poor wretch flees. I am not making chatter; I mean to change his life.*
>
> — *Annie Dillard,* Pilgrim at Tinker Creek

There is a double waterfall near the Rara Avis lodge and restaurant area. There is an overlook, the "mirador," with a spectacular view. There is also another waterfall associated with the property. This one is at the end of a several-mile hike on the Platanilla trail ending just outside of the reserve boundary and within Braulio Carrillo National Park. My wife and I hiked there many years ago, but the hike had been rushed. We raced out there to see the falls and then had to return for lunch to meet our children. This time, I am bringing my lunch with me. I have all day.

Another reason I want to hike this trail is because I know the area to be pristine. The trail, long and little-traveled, accesses the furthest reaches of the land preserved by Rara Avis. I know I will not see anyone and suspect it might have been weeks if not months since others hiked it. I recall a previous trip with Wilbur, the guide at the time, when I had glimpses of a Green Shrike-vireo and an Ocellated Antbird—two birds I hope to see again, only better.

Another reason this trail is little-traveled is because it is so wet. Heavily forested and dark, the trail begins behind the restaurant, near the dump, and descends for quite some time.

The original trail-building at Rara Avis must have been a frightful job—hacking a trail out of dense jungle. Shelves were cut into the mud in many places, and flat slices of tree trunk were used as steps. These are called "monedas" or coins. Perhaps they can be useful when there is adequate trail maintenance, but many of these have rotted, or worse, and have tilted downhill, providing a surface for slipping. It is slow going, because the mud-holes that aren't deep enough to fill your boots with water are still deep enough to suck them off your feet.

I slog along, listening and watching, and mostly thinking about my trip goal, seeing antpittas. These very characteristic tropical birds are sometimes described as a kiwi fruit walking on toothpicks. They have an oblong body shape and hop about on relatively long legs for their body size. They are reluctant to fly, but when they do, they accelerate like rockets. Living on the forest floor, they are typically some shade of brown. Many are not particularly rare, but their coloration and lifestyle make them very difficult to see. Rara Avis has two species that are relatively common, the Thicket Antpitta and the Black-crowned Antpitta. I would be thrilled to see either.

Not surprisingly, birds that are so hard to see often have loud distinctive calls in order to communicate with others of their species. For the most part, these calls are easy to learn, so I am listening for them.

On a previous trip, I had learned to my chagrin that antpittas do not always sing. I had come prepared to hunt for them. I had spent hours practicing before the trip to ensure that I would recognize the calls. This was a brief trip accompanied by my wife and a friend. I went birding all day and I heard nothing. I asked the guide and workers at that time. "Sure," they said. "We hear them all the time. Only we haven't heard them for a couple of weeks because of the drought." "What drought?" I asked. Everything seemed wet to me. They informed me that it hadn't rained for a couple of weeks and the birds had all become quiet. I didn't hear a single antpitta in four days. Consequently, one of the reasons for this ten-day trip was to give me plenty of time. I could wait for rains to come if necessary.

Fortunately, it has been raining. Birds are calling. I heard Thicket Antpittas while hiking with Giancarlo the day before. On this day, within a few minutes of being on the trail, I hear another Thicket Antpitta call in the distance. It is far off, no chance of seeing it. Some minutes later I hear a Black-crowned Antpitta. These are my favorite. I can describe their call as stunning, ethereal, mysterious. Here's the prosaic description used by Stiles and Skutch: "Song a very long, rapid series of short, clear, 'pinking' whistles, powerful and mellow, lasting a minute or more, starting very rapidly and gradually decelerating throughout." Think of it. This isn't a simple, recognizable backyard whistle. A minute or more is a long time for a single bird call. This call, to me, embodies all the mysteries of the rainforest. A dark denizen of the understory announcing its presence, secure in knowing it cannot be found. The call goes on so long, it has a "come-hither" effect, but I know if I head toward the sound, the call will stop, and the bird will slowly move off while staying in the thick vegetation.

My affair with the Black-crowned Antpitta began more than twenty-five years ago on my first visit to Rara Avis, when I knew next to nothing about tropical birds. On the hike into the lodge from Plástico, I encountered a young couple crouched on the trail. I stopped. They were birders. What could they be seeing? The young man didn't look pleased to see me. He turned to his girlfriend and said, "Shall we tell him what we've got?" Without waiting for an answer, he said, "It's a Black-crowned Antpitta. This is a very good bird." Yes, birders often use the adjective "good" to describe a very rare or hard-to-see species. I have to say, though, I'm not sure what constitutes a "bad" bird. Anyway, at that time, I didn't know anything about antpittas, but I took the explanation as an invitation to stay close and be quiet. Using their equipment, they recorded the antpitta and played the call back. The antpitta wasn't having it and moved off and became silent. In retrospect, I've learned that almost always, three people simply can't be sufficiently quiet to see this rare bird. The couple were both graduate students from Cornell, perhaps the world's most prestigious ornithological institution. They had expensive recording equipment and were on a quest to return with a collection of recordings of rainforest birds.

I did some reading and learned that this antpitta is found only from Costa Rica to Northern Colombia. It lives in a narrow elevational band, mostly to about 400 meters. Mid-elevations have been especially de-forested, so although locally common, this bird is not widespread. Recognizing the call of a Black-crowned Antpitta now, I realize how pleased I am to be back here, twenty-five years later, much more knowledgeable and with the time and experience, I hope, to see one of these rare birds.

I play my own recording and hear the antpitta again, but farther away. It doesn't sound very close, so I move on. It sings again in the distance. For now, I am happy the antpittas are calling.

A few minutes later, I hear a very raucous call. I am thinking *Red-throated Caracara!*—the exceedingly rare bird I spent yesterday searching for. Excitedly, I pull out my player again and play the call myself. Back and forth the two of us call. *I'm going to see a very rare bird,* I am thinking. Then it is overhead. It is all green, not red-throated. It isn't a Red-throated Caracara, rather, a Red-lored Parrot. I've mistaken the "k'chow" notes of the parrot for the "ca-ca-ow" notes of the caracara. *Some accomplished bird guide is laughing at me for this,* I think. Even though I am alone and probably a mile or more from another person, I am embarrassed. I said I wasn't good at calls, and the more raucous they are, the more they sound the same. I get over the disappointment of not having a Red-throated Caracara and am happy to have seen this common parrot.

I walk on—remembering. I have done the first part of this hike several times, and even though the first time was twenty-five years ago, I remember them all. I cross a stream that appears familiar. I can't help but leave the trail a bit and head downstream. These are the small pools of my memory. My wife and I once had a good time here. I think of the Steve Miller Band's *Jungle Love.* Not this time. I miss my wife, feel guilty for a second, and am then grateful for how much she encouraged this trip. My rapidly revolving brain adjusts once again to think of Thoreau: "What business have I in the woods, if I am thinking of something out of the woods?" I return to the trail.

Before my arrival, I had read some critiques online that indicated trail maintenance had been neglected at Rara Avis. There is plenty of evidence besides the rotten and slippery "monedas." There are potholes where previously dug drainage channels have clogged. Now and then a fallen tree has to be stepped over. Unfortunately, the worst is yet to come.

I arrive at a chest-high pile of trees and brush and vines. Apparently, there has been a significant blow-down. The debris filled a small ravine and covered the trail for perhaps 50 meters. There is a small stream flowing underneath as the topography presented steep slopes above and below. Going around is not an option. What to do?

I am intent on this hike and half-way to my objective. A cross-country route would be steep, and I know from experience I would slip and fall multiple times. I also realize leaving the trail would be the best way to become lost. This isn't country where one can use a distant mountaintop or even the sun for navigation. The trees are too tall, and the vegetation is too thick.

I already know what it feels like to be lost in the jungle. On a previous visit, my family flew to Tortuguero—the large national park on Costa Rica's Atlantic coast. The park is mostly wetland with only a few islands amidst canals, creeks and lagoons. One afternoon, after lunch, I had free time before an afternoon boat ride on the canals. I was intrigued by the jungle behind our lodge, which was situated on its own small, coastal island. A local guide had told me there was a White-collared Manakin lek back there (This variety of manakin "dances" and jumps and snaps its wings to attract mates. It is exciting to watch.)

I decided to look for the lek and whatever else I could find. Besides, a light rain had started, and I recalled that a little rain might bring out more birds, as it often seemed to at Rara Avis. Insects were abundant, so, despite the heat and extreme humidity, I put on a long-sleeved shirt and rubber knee boots.

The lodge described this small area of jungle as having a well-marked half-hour trail but, as I was to learn, it wasn't well-marked when under water. I don't know if there were several trails or if the water flowing on trails and washing debris around obliterated it in places, but as the rain became heavy, I lost the trail.

At first, I was not worried. I thought I knew where the trail was and believed I could trust my sense of direction. I also felt safe because the tract was supposed to be so small. Of those assumptions, only the latter had some truth in it. It was raining hard now and had been raining hard the previous day. Most of the jungle floor was under water. I followed what I thought were paths until I found myself standing in an area where several faint paths appeared to converge. I walked on, thinking I was on a trail. I stepped into an area with some water obscuring what I supposed was the path and, instead, plunged thigh-deep into a hole. I struggled in the mud and water to extricate one leg onto a small log and then tried to lunge forward with my other leg. I slipped and fell in hip deep. I clenched my toes so I wouldn't lose my water-logged boots and slogged across the hole to the opposite side, where I could grab some tree roots and pull myself out.

Although mildly spooked, I was mostly feeling ok because "it was a small tract." I saw an area ahead, through the trees, where there was more light. I assumed it was the clearing and the lodge. I struggled through brush and shrubs, feet underwater, giving some thought to snakes and insects, but plowing ahead anyway. I crashed through the shrubs and saw not the clearing, but a shanty—about 150 meters away. The intervening "open" area appeared swampy, but there was a monoculture of large vines and roots. I thought I would try to walk across. On my first step, I slipped off a root and plunged in over my waist. I now realized it could take hours—if it were even possible—to cross that swamp. I slithered out and returned to the jungle.

Now I was fearful. I had made a serious error in thinking that clearing was the way out. I had plunged ahead without regard to the direction from which I had started, and I had no idea which way to go.

The deluge continued. My boots were full of water. I was scratched and bleeding in several places. Mud had gotten in my pants and shirt and was uncomfortable against my skin.

Fortunately, I was able to suppress the feeling of panic. I did what I should have done 20 minutes before: I sat down and considered my situation. I realized that the tract was an island, and if I could walk straight, I would have to find a main trail or water or an impassable swamp as I had just experienced. Once I reached a side, I could turn around and try a different direction. I always carry a compass, so I used it to walk a straight line. I went slowly and kept looking behind in hopes of remembering landmarks. Soon I came to an area I recognized. I pointed myself in a direction I thought likely and found the main trail. In thinking about it, the area was small enough that I was never in any real danger. Yet I had experienced complete disorientation.

That experience is in the forefront of my mind as I confront the tangles now in front of me at Rara Avis. I am not going to get lost! If I plow through the jumble in front of me, I wonder, will I even find the trail on the other side? I am not sure, but I am sure I'll be able to return to where I am standing, which means I can't lose the path back to the lodge.

That last sentence makes me thoughtful. I am worried about getting lost and, in this case, know I can control it. How different from the common belief that how our lives turn out is a result of the "path" we choose. As related previously, I grew up in circumstances where I was often admonished to "stay on the straight and narrow path." As if we had control!

So many things knock people off their "path." Economic forces and illness are just two that are often uncontrollable. And, for many, certain paths simply are not accessible. I suspect being the cook or a worker at Rara Avis is a decent job. It wouldn't have satisfied me. Growing up in Los Horquetas, what choices were available? Certainly, they were not the same as mine. In this region of high unemployment, having any job may be enough.

Being born into favorable circumstances is no guarantee either. It doesn't take long to learn how vulnerable we are. A middle-school classmate of mine died in a corn-picker accident. A high school basketball teammate died in the Viet Nam War. As an adult, I've lost friends to early onset Alzheimer's and blood clots, and family members to cancer. I almost died myself in an accident—on a path—where I knew where I was, where I knew what I was doing. That's why I embrace the Buddhist philosophy: "It's not personal; you aren't in control." Nothing one does is without risk. Safety is a relative term.

As I've walked this morning, I've frequently reflected on how alone I am. I told Giancarlo I would be back by 5 PM, but darkness is scarcely an hour later. By the time anyone came looking for me, it would be well after dark, and then it would be a couple of hours more until this location could be reached. I don't hesitate to proceed. I make note of the position of some trees behind me, write a description in my notebook and start to climb through the blowdown. Half-way through a voice inside exclaims, *This is really stupid!*

I have said little about what dangers exist in this rainforest—besides becoming lost. But there is another—snakes. Each year between 500 and 800 people are bitten by venomous snakes in Costa Rica. The fact that only five to ten result in death is of little solace to me. Costa Rican tourist information says no tourists have ever been bitten, but how many tourists crawl alone through vine tangles for an hour while being two hours away by foot plus four hours by vehicle from a paved road? I know from my reading that those annual snake bites occur almost exclusively to field workers hacking their way through fields and brush—not so different from what I am doing. I also know there are many venomous snakes in this forest. Before this trip is over, I will see two fer-de-lance (one of several common names for *Bothrops asper*). On previous trips I had seen several pit vipers, including eyelash vipers which like to curl up, very well camouflaged, in tangles just like this. One of my books notes that three-to-six Costa Ricans die each year from the bite of this snake—usually sustained while "walking through dense vegetation."

I become very careful, while realizing close attention might not be enough when everything I am worrying about is so adept at hiding. I shake branches to frighten or move what might be residing therein. I move cautiously, wondering if it would be better to be noisier and more disruptive. Even so, I fall a couple of feet down into the tangle more than once, because the branch I am using for leverage unexpectedly breaks. Nothing happens.

Relieved, I re-locate the trail—much fainter now. As my concern for snakes abates, I remain paranoid about becoming lost. I place branches at some of the faint turns and make notations in my notebook, just to be sure. I know this jungle would become instantly terrifying if I didn't know where I was.

The trail begins passing through beautiful large trees, and my worries leave me. If birds were calling as I fought my way through the tangle, I didn't hear them. Now I do. A couple of my favorite calls are emanating from the underbrush. Anyone who has ever heard a Nightingale Wren won't soon forget it. Here are what some experts say about this song: "Confident to hesitant, rising and falling series of whistles... *hee hoo, hee hoo, hoo hee hoo, ss hoo hee...or tee tee-tee-tee ssi tee tee-tee-tee-tee-tee tee tee-tee ssi*" (*A Guide to the Birds of Mexico and Northern Central America*, Howell and Webb 1995). Although hard to describe, the song, once heard, is unmistakable. It has also been noted that the pure tone of the Nightingale Wren's whistle helps transmit the song in an obstruction-filled habitat.

I also hear several Slaty-backed Nightingale-thrushes. Their song has a slow, somewhat squeaky quality reminiscent of the Black-faced Solitaire (the bird caged in our friends' neighborhood) that is so characteristic of the high cloud forest.

I am having a good time now. *What will I see or hear next?* It is a crashing noise of something running away. I hurry up the slight slope I am on but see nothing. Looking down, however, I see the tracks. I have just frightened a tapir. Maybe one of Miguel's descendants. As usual, the tapir wasn't seen, but the tracks in the mud are still filling with water, and there at my feet is a still steaming scat. This is fun.

I have now reached the limits of Rara Avis. There are a few, decades-old signs nailed to trees, showing the boundary where Braulio Carrillo National Park begins. It is a much farther walk from where I am to trails or civilization within the park. In fact, from here, if the terrain is even passable, I would probably end up on a mountaintop if I strayed into the park.

After a quick look at the waterfall way down below, I walk along the "Boundary Trail," or what is left of it. Considering there are so few visitors, that few of these even bother with this morning's connecting trail, and that those who do start probably stop at the blowdown, I realize no one is hiking the Boundary Trail. I follow it for a short time, but I am, honestly, spooked by how faint it is. I am never in any danger of losing it, but I know myself: What if I become interested in a bird and forget what I am doing? In some places, even a few minutes of movement off the trail would make it very difficult to re-find. So, I begin to retrace my steps, and then I see a pair of Great Curassows.

Not more birds to describe! But these, according to Stiles and Skutch are "very large and robust, with long tail and prominent crest of erectile, forward curled feathers." Yes, three feet tall! The male is glossy black and the female mostly brown, but both are regal in their bearing, seemingly crowned with the "erectile forward curled feathers" that stand-out like a punk rock star's Mohawk haircut.

This family of birds is much persecuted because they are good to eat. In all my early visits to Rara Avis fifteen and more years ago, I briefly saw one. On this trip, I see them every day. That's what happens when Great Curassows are no longer hunted.

I sit in a few places to soak in the environment and begin working my way back. I am happy to observe that I have been over-worried about the trail—easily finding my markings and remembering where to turn without looking at my notebook. I take a deep breath and once again cross the tangle-filled ravine—except for a long pause in the middle when a troop of monkeys passes in front.

I hear branches crashing in the trees in front of me. Suspecting monkeys, I watch and soon see several. These are White-faced Monkeys. They have seen me, which sets off wild squawking and gesticulating. They are outraged. Their screeching continues well after I have passed from sight. This is more evidence of the wildness of Rara Avis. These animals are not habituated to humans—not like the ones that steal food off your plate at some of Costa Rica's seaside resorts.

White-faced Monkey

I begin to cross the various creeks. At one, I spy a Blue Morpho Butterfly fluttering past. They often fly at eye level and the color is at first startling, then mesmerizing, and then gone. I like butterflies. I know the difference between a Fritillary and a Satyr back home in Colorado, but, just as with some of the tropical birds, we North Americans experience nothing like Blue Morphos with their electric blue iridescence and five-inch and more wingspans. They are easy enough to film. Their flight is languorous. They float and bounce leisurely along jungle air currents, hypnotically flashing *blue blue* with every wing flap. Obtaining a photograph, however, is very difficult. The underside of their wings is a camouflage, dull-brown, and when one alights, that's all you see, if you can find it.

Even now, in my mind's eye, I can see those flashes of blue in the dappled light. Much as some bird songs and other jungle sounds evoke a sense of place, memory of morphos does the same. Morphos are widespread, but they do not thrive in city parks and gardens. They are one more of the many indicators of rainforest health.

The morpho's loud color does not translate to sound. The rainforest can be loud, as when a Crested Guan blasts its air-horn-like call at dusk, or when the squawks of a Barred Forest-falcon announce the dawn. What's remarkable is how quiet it can be. Many times I've sat and strained and heard nothing. Seen nothing. I've learned that often the best way to see something interesting is simply to sit as still as possible for as long as possible. While doing so, I'm often startled by a sudden movement. I strain to move slowly so as not to frighten whatever creature has entered my space. I crane my neck and I see a single leaf spinning. All else is stillness. I haven't heard tales of rainforest leprechauns or other spirits, but I've seen this phenomenon often enough to make me wonder. I suppose a random air current is to blame.

Commonly, nothing else happens, but on a regular basis, there will be an interesting sight. A few years ago, in a forest reserve north of Rara Avis, I saw a tapir trot into view and watched it for nearly a minute.

At the next creek, I think I hear a Dull-mantled Antbird—a bird I've never seen, so I have spent some time learning the call. I play my recording. There is a response. And then I realize the call has some similarities to a Buff-rumped Warbler. Again, there's a bird guide out there laughing at me. Still, I enjoy watching the small brown warblers with the relatively large buff patch on their back and rump. These birds also have a habit of swishing their tail back and forth, such that the light-colored patch on the rump is in constant motion—probably an adaptation to confuse predators.

Now, the rain that has been threatening all morning begins. I have been walking in drizzle for some time and have already realized my first packing mistake. Temperatures at Rara Avis are always in the 70s. What good is a raincoat at that temperature—especially combined with the exertion of hiking? Inside a raincoat, one is hot, sweaty, and wet—just the same as being in the rain. I recall that on a previous trip I made a mental note to bring a small umbrella. I had forgotten that mental note and am essentially at the mercy of the downpour. I am also hungry. I look around and notice a small island. On the island is a large-leafed plant (*gunnera)*—known colloquially as "sombra de pobre," or "shade of the poor." Thus, it might serve as a rain-shield for the poorly prepared. I hop from rock-to-rock out to the island and find a comfortable seat under one of the large leaves. While sitting, the leaf is several inches above my head and wide enough to provide complete protection from the rain, and, better yet, I can still look around.

I had so longed for this trip, and here, on my first full day alone, I am having a fantasy lunch—sitting mid-stream, heavy rain falling, but dry because I am taking advantage of what little local knowledge I possess. I open my package of cold, refried beans, spread them on some crackers, munch a carrot I scrounged from the Rara Avis kitchen, and I am very happy.

Every heavy rain I have encountered in this environment lacked accompanying wind. This rain too comes straight down and heavy. I begin to watch the water level in case there is a flood. There will not be one today. Sadly, two tourists once drowned in a flood while swimming in the pools below the waterfalls near the restaurant. They had ignored signs proclaiming the danger of swimming in the afternoon when a storm upslope might send a major flood without any rain falling on Rara Avis itself. I myself know how difficult those inviting pools are to resist when rain isn't present and doesn't seem imminent. Since then, measures have been attempted, including an alarm to alert swimmers of rising water levels, so thoughts of floods were no joke.

We were once here for a big flood. My family and I were in the restaurant for lunch. We had noted on the way in that water levels were high. It had rained during the night, and it was still raining in the morning. Then it began to rain harder. We decided to make our way over to the double waterfall nearby.

There are two falls near the lodge. The first is approximately ten meters in height, the second about four higher. Between the falls is a nice swimming hole, about 15 meters in diameter, then there is a ledge and the second falls begins. The trail for viewing the falls hangs on the side of the valley through which the creek flows. The falls cannot be viewed from above on the lodge-side of the creek. Several of us worked our way down the rough and wet trail until we were about five meters below the top of the upper falls and about ten meters from the water. The water was so high and came over the upper falls with such momentum that the two cataracts had converged. The sight and sound were overwhelming. The roar was deafening, and the power of the water was terrifying. After taking it in for a minute or so, we slowly backed away shaking our heads. There was nothing to say.

There is a bridge over the creek above the falls, where I estimated the water had risen at least two meters. But most interesting to me was the scene the next morning. It was impossible to tell there had been a flood. Probably, some boulders had moved, but for all general appearances, the stream looked exactly as it had before. This was a lesson to me that I've recounted tiresomely to others.

Streams evolve their shape based on the timing and amounts of natural rainfall. Under such circumstances, watersheds are remarkably resilient. However, once we remove water for irrigation, cut trees for development, shorten meanders to build a road, or undertake any of the myriad other reasons humans have for altering the landscape, then the watershed can no longer cope. Landslides from heavy rainfall typically are not "natural" disasters, but the logical consequence of poor watershed management and bad engineering. This was obvious when we left Rara Avis on the visit of the huge flood. We saw the devastation to the road and the overcut areas below the protection of Braulio Carrillo National Park. It makes me crazy when I read of a project to straighten or dam a stream. The news item always says something like, "Environmentalists worry that the project may affect the watershed and associated wildlife."

There is no *may* about it; you mess with the watershed, and there *will* be effects.

Once again, Thoreau comes to my rescue. I recall the admonition I quoted previously, and I stop thinking of watershed disasters and begin to look around again. I can sense the daytime jungle "calling it a day." Sunset is brief enough and on dark afternoons, the days seem to end particularly early. A heavy fog/mist is rolling in as the rain slackens. This is a perfect time to finish my hike. I am exhilarated, even though the only sound is the sucking of each boot as I pull it from the mud.

Twilight finds me on the small deck outside my room. I have lugged in a box of wine and have a glass in hand. I sip my wine in the gathering mist, reminiscing about previous visits, and I recall that near dusk, the rare Short-tailed Nighthawk often flies along the stream. A good place to see one is near the bridge. I rouse myself, walk down the stairs and stumble down the small hill toward the bridge. Usually, I walk down the concrete steps toward the restaurant and make a sharp left and walk along some monedas toward the bridge, making a large "L" with my path. Instead, I take a short-cut, walking through the vegetation toward the bridge in a straight line. A large shape rears up almost in my face, long, brown wings rocking as the bird disappears behind some trees. "A Pauraque," I think. I heard their "purr-WEEE-eer" calls both previous evenings.

Waiting, I suddenly see it flutter back into view and disappear on the ground. The camouflage of these birds is astonishing. I make a movement toward where I saw the bird land. I can't see it, but it can see me, and it abruptly rises again. Noting where the bird had seemed to emanate from bare soil, I creep closer and wait. It returns, but I still can't see the bird. After a while, I finally spy it sitting quietly and all but invisible against the brown leaves and mud. Another brief movement from me and the bird flutters off once more. This time I approach even more carefully and find what I am looking for, two speckled brown eggs deposited neatly on some leaves. I look about and look back. No eggs! Yet, they are still there. I find them again eventually, but what masters of concealment these birds are. I back away and give the area wide berth for the rest of my visit. What a great day!

CHAPTER 5

WINNING THE FINAL FOUR

For that forty minutes last night I was as purely sensitive and mute as a photographic plate; I received impressions, but I did not print out captions. My own self-awareness had disappeared; it seems now almost as though, had I been wired to electrodes, my EEG would have been flat. I have done this sort of thing so often that I have lost self-consciousness about moving slowly and halting suddenly. And I have often noticed that even a few minutes of this self-forgetfulness is tremendously invigorating.

— *Annie Dillard,* Pilgrim at Tinker Creek

Science is not only a disciple of reason, but, also, one of romance and passion

— Stephen Hawking

On March 31, 1997, Miles Simon and Mike Bibby led Tucson's University of Arizona Wildcats to the NCAA basketball championship. I was in a hotel room in Maryland, on business travel. That was difficult. I wanted to watch the game with my son—an equally rabid fan. I leaped repeatedly off the floor as the game ended. I had to control it and jump softly; there were people in rooms all around who weren't sharing my emotion. For several minutes, I could barely contain myself. If I had been home, I would have been screaming. My heart was racing. It was challenging to be that exultant and to have no one to share it with. (Even if you aren't a sports fan, you know someone who is, and how they can become so inexplicably excited. That was me!) April 6, 2013, I had the same feelings. I saw a Thicket Antpitta. I did it all by myself.

Now, I look back on the passion I felt on those occasions with some mixed feelings. I still revel in the jubilance I felt, especially the fact that I could do it. But there is some sadness mixed-in. I grew up in a home where my earliest memories are tinged by my mother preaching "self-control, self-control." As I continue looking back, I recall her remarks when as an eight-year old I hit a home run, and rather than celebrate with me, she commented on the amount of emotion I displayed. Even today, I can feel that eight-year old's shame for being emotional.

Outwardly, I had a very successful childhood and adolescence. My hometown of Highland, Illinois was a very small pond. I could be one of the bigger fish: excellent student, athlete-of-the-year, Homecoming King, etc. However, achievements were always colored by a sense that I still wasn't good enough, or that I shouldn't forget to dwell on the idea that I was a "sinner in the hand of an angry god."

Indeed, one of my mother's favorite sayings, uttered when we would see someone who was crippled or poor, or if we read of a tragic death, was "there but for the grace of god," meaning that I should never exult about anything. I know now she didn't intend for it to affect me as it did, but I felt guilty about anything I achieved and I lived in constant fear of losing anything I cared about. Her other favorite saying, spoken whenever she was in a bad mood (which, sadly, was often), was, "It's a great life if you don't weaken," which I understood to mean, "Never drop your guard or show emotion." The two things missing from my childhood were compassion and empathy. If no one is giving those to you, how can one feel the joys of passion?

Mom was bitter about her childhood, but in our keep-a-tight-upper-lip, stoic German community, that was something you could not acknowledge. Being forced to hide her highly emotional personality, Mom molded herself into an outward existence of being guarded, stifled and uptight.

She frequently complained bitterly about her father's favoritism of her siblings. She reminded me relentlessly, whenever I complained about school or working, how she was forced to quit school and clean chicken coops when she was fifteen. It was only later, as my education ensued, that I recognized her raw intelligence and what shackles her inability to be educated had put upon her.

Her father sometimes abused alcohol. He was functional enough but often surly and mean. She turned 15 early in World War II. Her brother, who returned from combat with what we now recognize as PTSD, was already serving. Those were fearful times. No wonder she had to go to work, and no wonder that her younger sister, born into better times, was welcomed as a joy and was coddled.

While I know nothing of my uncle as a youth, I know that he never functioned normally in the world after the war. Time after time, when he'd have a breakdown, or leave/escape from an institution where he was living (I'm not sure which is the appropriate word...Mom wouldn't discuss it), it fell to her to rescue him. I recall a Christmas Eve when this happened.

Let me set the stage. My dad had a small shoe store with no employees. He worked six days and one evening per week. There were no vacations—none. His help in those early days, before I could work, was Mom. On Christmas Eve he closed the store at 4:30. This meant one hour less work, followed by an entire day off. How important must that have been! A call came about 5 PM, and Mom had to leave with an ambulance to collect Uncle Darrel near Chicago, four hours away. I can still see my mother. I was probably 10 or 11, and, at the time, didn't dwell subjectively on what I was seeing. The veins in her throat, nearly popping, and her red face are my objective memories. She stifled whatever she was feeling, but it was clear she was taking on this burden with a controlled fury.

She was back by the time I awoke on Christmas morning, sleepless and tight-lipped and angry at life for having done that to her. I later learned she was also angry at her father and her siblings for, in her mind, never helping. Her controlled rage and disappointment in life were always close by.

Accordingly, I was fearful of my mother. We showed little physical affection—no kisses, only awkward hugs. My wife and friends would probably find it strange to hear what she once told me: "You were always a 'touch-me-not.'"

It was many years before I finally understood that she was also afraid of me. It is poignant for me to recall her telling me that once, as an infant, I was crying uncontrollably. She said, "I just laid you on the bed and you kept screaming. I just didn't know what to do." That scene encapsulates our relationship. I now recognize that no one could have loved me more. She was full of compassion and empathy but had been taught not to show it. Her life had been shaped by alcohol, war, and her mother's piety for Catholicism. There was no place for her own emotional life. It was no wonder she impressed the same things on me.

Fortunately, there was one big positive. The stifling of her natural brilliance was so frustrating that it came out in her continual support of my reading and learning. When I was perhaps nine years old, she saw I was bringing books home from the library. She talked to the librarian and had her oversee a booklist for me. I don't think Mom knew Hawthorne from Dickens, but she knew instinctively that if I was going to read, I needed to get the most out of it. I credit the wide range of books I read as a child, everything from Bullfinch's *Age of Fable* to the *Arabian Nights* to the aforementioned Hawthorne, with instilling in me a wide-ranging curiosity and belief that almost anything I pick up and read will have something of value in it. I'm so grateful for that. I'm grateful for it because it eventually solved the problem of my mother and me.

Much reading, some counseling, and plenty of contemplation led me through a variety of mind-states. Early in my adulthood, I was still fearful of my mother, trying to win her approval. Sadly, she died tragically young before I fully grasped who she was. It was a long time until I understood that she could not show compassion and empathy because of the difficulty of her childhood.

I acknowledge that everyone's personal emotional make-up is different. We don't all perceive the same circumstances in the same way. Simply being a year or two more-or-less mature in emotional development can make all the difference in how a childhood is perceived. What Mom experienced growing up says little or nothing about what her siblings experienced. The same is true for myself and my siblings. For me, I grew up with shame and fear, but, having been dealt a very good hand, I am pleased to say my feelings toward my Mom are now empathy and compassion. How I wish she could have lived to see how well her children have done! I believe she would have relaxed. I believe she would have enjoyed seeing passion in me and maybe allowed some in herself.

As for me, I was fortunate to find an understanding partner whose encouragement and acceptance immediately overcame any residual reticence I may have had. Indeed, even now, she recognizes my passion for being outdoors and enjoying wildlife. It was she who encouraged me to take this trip and shushed away any reluctance or guilt I might have felt for going off alone to chase antpittas.

My search of antpittas only proves the old saying that "there is no accounting for taste." Most who read this probably have never heard of them.

Antpittas are members of the family *grallaridae*. Although common-to-abundant in much of Central and South America, they have no counterparts north of Southern Mexico. Antpittas can be broadly lumped with a large group of species sharing typical characteristics of drab colors and retiring habits. It occurs to me as I re-read the last sentence that I could be talking about rodents or even insects. No, they are birds! As with the many birds with "ant" in their names (e.g. antshrikes, antwrens, antvireos, antthrushes), ants aren't typically part of their diet, but a reason this group of species has no direct analogs in the United States is that they are essentially solely insectivorous.

What explains my love for antpittas and their kin? Certainly, some of it comes from that first encounter with the researchers from Cornell on my first trip to Rara Avis. My interest was kindled more after I organized a few group birding trips led by some phenomenal bird guides. When I asked, they said antpittas are almost never seen with a group. "They are too shy." Now that I'm more experienced, I suggest one's chance of seeing them decreases at least by the square of the number of people trying to glimpse one—unless you have happened on an antswarm.

To me, antswarms had seemed as mythical as antpittas. I remember how envious I felt, when having a beer in a bar near Carara National Park in Costa Rica, I overheard several birders excitedly discussing the antswarm they had encountered a few hours before. They described a Streak-chested Antpitta that had come within a few feet while feeding.

What exactly is an antswarm? It is NOT what is depicted in the old movie *The Naked Jungle*, starring Charlton Heston. In that movie, army ants are a scourge that will devour humans, animals, anything in their path. The reality is much different. My ornithological hero, Alexander Skutch, used to welcome an antswarm into his home in the rainforest, because once they had passed through, most of the vermin that had collected had been cleaned out.

An antswarm also has no resemblance to a collection of ants foraging in your home. For example, my wife Mary is almost perfect. But one characteristic I would change is to make her reactions to unpleasant events commensurate with their gravity. Her scream sounds the same to me whether she believes my habit of birding while driving is about to cause a head-on or whether she's spilled some water on the floor. On a recent trip, we were renting a small house near San Vito in southern Costa Rica. She was preparing dinner when I heard her scream. *She's cut or burned herself badly,* I thought, and ran to the room. No. It was ants! The *pan pequeños* we had picked out for breakfast were literally black with tiny ants. The sack containing a whole-grain bread, difficult to find here, had also been breached. A heavily covered poppy seed roll was never so black. The pan pequeños were history. But we brushed off the bread and re-wrapped it. The ants were unpleasant, but not a catastrophe. The next morning, we went birding at the nearby Las Cruces Biological Station (aka Wilson Botanical Garden)—part of the Organization for Tropical Studies. We were going birding, but what we really wanted to find were swarming ants! So why didn't we stay home and watch our baked goods?

Eciton burchellii are popularly known as army ants. Peaceful Costa Ricans love to point out that such ants are the only army in their country. It is true that they are a swarming army of sorts. Nearly 600 species of birds and insects have been documented to associate with army ants, and some are wholly dependent on them.

Army ants are nomads. They form armies of half a million or so and forage from temporary nests or 'bivouacs.' For three weeks, they march out and return to the same camp until the larvae hatch. Then for two weeks, the army moves to a new location every night. Finally, the larvae pupate, and the ants select another three-week encampment. These ants forage as an army spreading out over the jungle floor, crawling up limbs and tree trunks and hunting under every leaf. Insects of every kind flee frantically as they are disturbed, and that's what brings the antbirds.

Antbirds, including antshrikes, antpittas, antvireos and more, are found only in the lowland tropics. While not all of these depend on army ants, some are obligate ant followers, meaning they forage not on the army ants themselves, but nearly-exclusively on the insects fleeing the army ants. When a large group of birds is attending foraging army ants, the event is called an antswarm. An antswarm can have more than a dozen bird species with multiple individuals of each type. It is a feeding frenzy, with usually shy birds landing almost at your feet.

Most of the antbirds, particularly those that forage with army ants, are shy and difficult to see, unless they are attending an antswarm. Many have habits more like forest rodents, keeping quiet and close to the ground. All of them need large areas of mature forest. Despite many previous trips to the tropics, whether it was due to bad luck, being with too many other birders or being in the wrong location, I had never watched an antswarm. I still had never seen several of Costa Rica's antswarm obligates and had only fleeting glimpses of others. I wanted to see the new birds. I wanted long and satisfying views of the others. I wanted to experience this phenomenon that I had read and heard about for so many years.

We did find two small antswarms at the Las Cruces Biological Station. I recognized one of them when I saw three or four Swainson's Thrushes leaping up and down from near a fallen log. As I described earlier, Swainson's Thrushes are known to North Americans for their flute-like songs delivered in the shady mixed coniferous forests in which they nest. This was late April, the peak of their migration through Costa Rica, and they were acting like antbirds. Where were the antbirds? I could hear other individuals, and eventually I spied two species I had never seen before. But I wasn't satisfied. There were not enough species or individuals. The birds were still shy and sort of dropped in and out. *Is that all there is?* I wondered.

Two weeks later, at a more remote location, this time in Northeast Costa Rica, I finally realized my dream of encountering a king-sized, fully raucous, wildly entertaining antswarm. I heard the feeding call of a Bicolored Antbird. I looked about and saw army ants all over the trail. I found a good vantage point next to a tree and watched. An Ocellated Antbird appeared, and then another. Then a Ruddy Woodcreeper dove in, snatched a large insect, and clung to a nearby stem while consuming it. Next came two more Ruddy Woodcreepers, two more kinds of antbirds, and three more species of Woodcreepers. Eventually I tallied 12 species. I watched the spectacle for more than an hour. I was thrilled. I have been fortunate to have had many noteworthy wildlife experiences, and this one was as good as any! I guess my wife and I make a good pair—both of us have emotional reactions to ants.

And Costa Rica is home to a lot of ants—more than 1000 species. That number is approximate, because it is certain that not all have been described. The great naturalist E.O. Wilson, who spent much of his life performing research on ants, said all he needed was an hour or two in a tropical rainforest, and he would be able to find a species previously unknown to science. This, of course, is one of the oft-repeated reasons for protecting the tropics, that we don't know what treasure trove of natural medicines are yet to be discovered, and that we might drive to extinction without even knowing them. Wilson himself has said, "The one process now going on that will take millions of years to correct is the loss of genetic and species diversity by the destruction of natural habitats. This is the folly our descendants are least likely to forgive us."

More familiar to tropical visitors than army ants are *Atta* ants or leaf cutters. (*Atta* is a genus of New World ants of the subfamily *Myrmicinae*, containing at least 17 known species.) Walk almost anywhere in a forest and you will encounter a trail of ants all clutching a part of a leaf on their back. It doesn't quite fit Macbeth's utterance that he would not be fearful until "Birnam wood remove to Dunsinane," because whole trees are not moving, but don't underestimate the power of these ants. Leafcutters are the dominant herbivores of the New World tropics. The amount of vegetation cut from tropical forests by the *Atta* ants alone has been estimated at 12–17 percent of all leaf production.

Atta ants, with their wobbly and lengthy columns, are fun to watch and benign, as long as you aren't a favored tree. But then there is the Bullet Ant, so named because the sting is reputed to feel like a bullet wound. These large ants (2.5 centimeters or one-inch) will bite hard and then twist their abdomens to deliver a sting. Some years ago, on another visit to Rara Avis, a worker was bitten by a Bullet Ant. The worker toughed it out and stayed with his companions, but whenever I saw him, I was struck by his flushed, red face, and watery eyes. He looked beyond miserable. Supposedly the pain lasts no more than 24 hours and begins to subside after eight or so.

I am glad utilitarian arguments can be made for saving biodiversity, but such discourse makes me uncomfortable. *Atta* ants and their little green trains are amazing as is the one-inch Bullet Ant that can sting a 200-pound human into submission. They should continue to exist for their own sakes and be protected for their intrinsic value as products of millions of years of evolution. Each, in its own way, is a work-of-art. Besides, why should we deprive our heirs of seeing them—or, perhaps more important, seeing what they may become after many more years of evolving?

I feel frustration because too many humans have so little regard for our fellow creatures. I once sat through a commencement ceremony at a tiny rural western high school. One of the "cowboy poets" was a featured speaker. The audience delighted in his jokes about cooking and eating endangered species. As the controversy has continued, I see more clearly that resentment has arisen from rural folks with low incomes feeling that others (elites like me), with income not reliant on the land, are telling them what to do. It is a problem already too late in its solution. For me, I believe the solution lies in those of us who do have means providing the tax revenue for jobs that maintain endangered species. Whether it be clean energy or more enforcement personnel on public lands or resurfacing roads, repairing trails, and re-building bridges, we all must contribute. The difficulty is when government leadership is lacking. That makes it tough.

My idea, you see, is that when there has been government leadership to give everyone a job, those societies not only thrived, they've left many monuments for their ancestors to marvel at. My belief is that there were not unemployed individuals in ancient Egypt when the pyramids were being built. There were not idlers, living off the state, when the Mayans built their monuments. No doubt many of the workers who built those edifices and the like were slaves or nearly so. But it doesn't have to be that way. Let our government build things now by paying a good wage. Keep everyone busy and build things of real value.

Personal examples include the beautiful stone buildings at some state parks in Illinois near my birthplace (Pere Marquette, Giant City). Here in Colorado, the most popular trail in our Colorado National Monument is the Serpent's Trail. Who built them? The CCC (Civilian Conservation Corps)! I knew conservative families growing up who still groused about what a waste was The New Deal, calling the WPA (Work Projects Administration), "We Piddled Around." I'm sure "piddling" occurred; it still occurs everywhere. But much was also accomplished. It is impressive how many of the New Deal-supported buildings, roads and trails are still in use, with many having become national historic sites.

Unfortunately, the conservative side of our government sees only one way that is feasible or worthwhile for government-funded jobs, and that's with military activity. There's so much we could do with a similar civilian conservation program today, such as repairing park infrastructure, restoring streams and more.

Enough! My mind has wandered far from my antswarm encounters, and there is a critical detail I haven't mentioned. None of the antswarms I watched were attended by antpittas. Both locations were at elevations where antpittas are rare. For myself, I can note that in a couple of collective weeks of birding in those areas, I've never heard an antpitta call. Eventually, I realized if I was going to be lucky enough to see an antpitta on my own, I needed to go where they were common, and people were not. That's why I am here at Rara Avis.

Now that it is my third full day, I am losing some confidence. I don't seem to have the skill or the luck. I reflect on a memory of a lodge where Scaled Antpittas were sometimes observed in the adjacent forest. I know this because a guide I trusted told me he had heard one.

I was to learn, however, that some guides ignore antpittas. One night at dinner at this lodge, there was a guided group of photographers. They were all excited about the birds they had seen and photographed on their trip. After talking to a few of them, I knew they were primarily on a birding trip and that they had, indeed, seen some interesting species. Thinking that their guide, who had told me he often came to this lodge, might know something about the local antpittas, I interjected myself into their dinner conversation and asked him where he had heard or seen antpittas. His response surprised me. "You mean the small brown birds on the ground that no one ever sees?" "Yes," I said. He continued, "I don't even bother with them. It takes too long and you usually don't see them anyway." He had no idea which antpittas, if any, were found in the surrounding forest.

At that lodge, I started before dawn and slowly walked the trails that seemed to have the right kind of habitat. I neither saw nor heard an antpitta, but then I met my wife for breakfast. "I was looking for you," she said. "I saw this neat bird. It was oblong and had two skinny legs and hopped down the trail when I was hiking." I asked her a few more questions, and she perfectly described a Scaled Antpitta. I suppose Mary's good luck, and my bad, further fueled my desire to somehow see an antpitta.

The difficulty of seeing these birds also was firmly imprinted on me during my first trip to Ecuador. We were in Amazonia on the Río Napo, upstream from the city of Coca. The lodge was owned and operated by an indigenous Quichua community. The guide was a local youth. I was never sure exactly what his nickname was, but it sounded like "Chiri." Chiri could mimic most of the sounds of the birds and could move through the forest so silently, it was as if we were accompanying a jungle creature. In this pristine forest, he was detecting birds that none of us would have seen. At one point, we were trying to see one of the antbirds, but none of us could follow its movement except Chiri. Eventually, I realized the bird was essentially bouncing off the branches—especially the one it had left prior to assuming a quiet perch. I was focusing my binoculars on the waving branches and having an excellent view of where the bird had just been. With Chiri's help I finally saw the bird. I realized nearly all people walking through that forest would never see a bird that acted like this. They would see movement and no more. I suspect the ability to leave a perch noisily and land stealthily was a behavior learned to deter predators.

So often that had been my story—seeing only movement and nothing else. Without Chiri to help me see what else was there, I would have missed a lot. This went on all morning with Chiri pointing out various birds, plants and insects. I had sort of a "eureka moment" when I associated the jungle understory with the ocean and a coral reef. Viewing the ocean over a coral reef, snorkeling, reveals little about what is below. But when you scuba-dive "within" the reef, you see that every little crevice holds some creature impossible to see from above. In the same way, the jungle understory was a thriving and amazing world equally unknown and unseen to those taking only a superficial view. Unfortunately, the comparison ends there. With the right equipment, many inhabitants of coral reefs are easily seen, whereas seeing antpittas and their kin typically requires lots of time, experience and luck. Even with Chiri, we saw no antpittas—probably because there were seven of us.

I am thinking of the skilled Chiri who couldn't show me an antpitta. I am thinking of my wife who had more luck. Maybe this trip wouldn't yield my hoped-for payoff either.

In preparation, I had reviewed both the "bible" of Costa Rican bird watchers—*Birds of Costa Rica*, by Stiles and Skutch, and the more recent field guide of choice, by Garrigues and Dean. Of the 26 Costa Rican species with "ant" in their name, only three are called uncommon or rare. For the Thicket Antpitta, it is noted in Stiles and Skutch that it is "widespread, locally-common." Unfortunately, the book also notes that the bird is "seldom seen, its abundance is indicated by the often-heard song." The latter statement certainly didn't give me confidence that I would see one.

I was revising this chapter in mid-February of 2018. The previous thirty days are probably the most heavily birded period of heavily birded Costa Rica. I checked the electronic database ebird.org and in the past month, there had been only 23 recorded locations with recent sightings of Thicket Antpittas, and several of these sightings were from the same person. In addition, the database does not require one to note whether the bird was heard or actually seen, meaning many of the sightings were more likely "detections" rather than visuals. Finally, the total number of checklists marking Thicket Antpitta was approximately 40 out of nearly 8000 since the beginning of the year.

I was shocked when I looked up Black-crowned Antpitta. There were zero sightings so far in 2018 and only four listed for the entire year 2017. The entire data base, from more than 150,000 checklists for Costa Rica, shows fewer than 100 total detections for Black-crowned Antpitta. Had I known it was this rare, would I have even tried for it? (For comparison, for the week starting on January 22, Lesson's Motmot had been recorded nearly 1,200 times—in a single week!) At least the guidebooks judged the Black-crowned as rare. Anyway, I was building my excuses and rationale for accepting failure.

So, I wonder. As described in Chapter 1, Lesson's Motmots can adapt. What about antpittas? Will they always be here? They are so shy. They won't fly across roads. They won't fly across pasture. As habitat becomes more and more fragmented, are they doomed to become extirpated patch-by-patch until none are left? I want my descendants to have the opportunity to see them. My children are not that interested in seeing birds, although both are such outdoor lovers that doubtless it would cause them pain to hear of a species' demise. But what about my grandchildren? Will they be able to see them if they want to?

Lots of things on my mind! Why are antpittas so hard to see? Well, unlike the more colorful manakins I've previously mentioned, antpittas eat mostly insects. Insects hide in deep, dark places because they don't want to be seen or eaten. Birds that frequent such areas don't want to be seen or eaten either and must be furtive to find and eat creatures with the same motivation.

Fruiteaters such as tanagers and manakins often associate with other species when feeding at a fruiting tree. With so many potential sentinels, one of the birds will usually spot a predator and give an alarm call such that all of them can flee or hide. Fruit-eating has another benefit. Fruits, unlike insects, want to be seen and eaten. They advertise their presence with bright colors and luscious odors. That's another reason many fruiteaters have developed their elaborate courting rituals. Dinner is often easy to find, leaving ample time to compete for sex. Not so for the antpittas who comb the dark understory for their meals, using stealth to find their prey. This also explains why many antbirds are paired throughout the year, indeed for lifetimes. There is no time for leks and flirting.

Thus, because I hope to encounter antbirds, I've learned to creep slowly and scan the trail and vegetation well ahead of where I am walking. It is why I have walked for miles deep in tropical forests. Sometimes I have only seen three or four species (usually, none of them antbirds) in several hours. As I said above, most antbirds won't fly over an open area. This is why, in some cases, there are distinct species, all with a common ancestor, on either side of the Amazon River as well as on large river islands. The only reason antpittas are not virtually undetectable is because of their calls. Because they live in the thick understory, they have calls with a relative low frequency (~2khz). Those are the frequencies that can best be heard through all that vegetation. These lower frequencies happen to be good for humans too—especially those, like myself, with high frequency hearing loss. (Another reason I am fond of antpittas!)

On my first full day here, Giancarlo and I had heard Thicket Antpittas several times. Once we tried to call one, but the bird was not close by, and it seemed to move away. I heard one on my hike on the Plantanilla Trail on Day 4. Once again, I had played the call a few times, but had no success. Antpittas seemed as far away and as discouraging to me as ever.

All this thinking is muddling my too-busy mind as I awake in the semi-darkness of dawn. My day begins with promise. Before breakfast, in the near darkness, I walk down to the river and across the bridge. I hear a soft *uhu*-oo. As I am thinking, *It might be a quail-dove*, a Purplish-backed Quail-dove strolls out from the understory. As said previously, Quail-doves have been referred to as "vampires of the rainforest," because they are always in the darkest areas. About the only way to see them is by slowing walking alone. Occasionally, I have been with groups, where the first person, typically the guide, catches a glimpse of one, and that's it. As soon as they sense someone's presence, quail-doves rotate as if on a swivel, and glide/walk, as if on wheels, into shaded vegetation, and vanish. If sufficiently startled, they flush with a flutter of wings. By the time I am aware of them, they are gone.

This time, I have a special view. The Purplish-backed Quail-dove walks into a patch lit by a shaft of the rising sun. The bird stands in the sun for a moment, possibly blinded by the unusually bright light. It does a pirouette. It is beautiful: slate, blue-gray with a prominent mustache. The back is a deep purplish-maroon. An amazing beginning for my day!

I stroll back to the buildings, elated, my heightened senses aware of my surroundings. High in the trees, I hear a frantic "eat-it, eat-it, eat-it, nowww," from a Bright-rumped Atilla. Near the riverside, there is also the explosive song of a Bay Wren followed by its annoyed chatter. How great it is to learn a few calls.

Following the usual breakfast of *gallo pinto,* I head up the Nicolas Trail. A King Vulture sails just over the canopy, affording me a better-than-usual view. Most of the time, they are sailing high overhead, easy enough to identify, but too far away to appreciate the creamy-white body and crimson, featherless head. The idea of a vulture-lifestyle tends to make us queasy, but the beauty of this bird belies its macabre eating habits.

A purpose for this morning is to see a White-crowned Manakin. Manakins are compact and chunky with no apparent neck.

They are adept at zipping about like tiny rockets, often bouncing off one branch to land on another, leaving the hopeful viewer checking out where the bird just was. As with hummingbirds, male manakins do not help with child-rearing. Not unlike some human males, their lives revolve around eating and then showing off with their buddies in hopes of attracting a female for a one-night (or less) stand.

This behavior, of hanging out in a group and showing off for prospective mates, is called "lekking." This species typically flutters "butterfly-like" among bare branches, with individual males displaying as much as 50 meters apart. As such, its behavior has not been studied as well as its cousin, the White-ruffed Manakin, whose males gather in groups and take turns dancing on a fallen log, making them much easier to see.

The ebird database for Costa Rica showed only six locations and less than ten sightings in the first six-weeks of 2018 for White-crowned Manakin. In contrast, not only are there dozens of locations where White-ruffed Manakins have been sighted, at many of these it appears the birds are encountered every day. I once encountered a female White-crowned, but until later that year, in Ecuador, I'd never seen a male.

I wanted to see a male, because this species, as with most manakins, is exceptionally sexually dimorphic. Males are velvety-jet black with an ivory-white crown. The females, who do all the work such as nest-building, brooding, and raising the young, are a dull, leaf-colored green. It has made me wonder if male manakins even know what their nests and young look like. Even in the non-breeding season, males are known to hang out together and practice their dances from time-to-time.

I had been told a few years ago where there was a lekking area so I thought I would check it out. Immediately, I see White-ruffed Manakins, but the White-crowned did not appear. Nonetheless, I am having a pleasant time listening to the forest. I am always within earshot, it seems, of a Nightingale Wren and a Slaty-backed Nightingale-thrush. Hearing their songs keeps me happy. The trail I am hiking eventually intersects the boundary trail I was on the day before. I pick a nice spot for lunch, and within minutes a female curassow walks by within a few meters.

Back on my feet, I hike the opposite direction from the previous day on the boundary trail, finding it in decent condition, until I come to a trail that follows the river back to the lodge area. This trail is badly washed out in places. At times I have to dig in with my boots and slide several feet down a muddy slope. Eventually, I come to a junction where the left fork seems to descend steeply toward what sounds like a raging torrent. I have been here on previous trips, but never descended, assuming the trail was too steep and that the stream could not be crossed because of its turbulence. This time I decide to take a closer look.

The trail descends easily to the riverbed and not to the steep overlook I expected. Here the stream is wide and shallow, unlike the cascades just above and below. Across the stream, I can see what appears to be the trail's continuance. There are plenty of boulders to use both for support and for steps. I am surprised how easily I cross.

Immediately, a fruiting tree catches my attention. I swing my binoculars and find a Swainson's Thrush, a Wood Thrush and the rare Blue and Gold Tanager. Swainson Thrushes and Wood Thrushes may co-occur somewhat during migration in Eastern North America, but they nest in different habitats. Wood Thrushes still nest wherever larger wooded areas remain near my hometown in Southern Illinois. I know them from my youth, but it never occurred to me that either species would spend more than half of their lives hanging out with this rare and beautiful tanager.

As I've said over and over, most of us in North America do not think about where the birds go when they "fly south for the winter." Instead we should be focused on it. "Our" birds need this habitat if they are going to survive. Sadly, the Wood Thrush population has declined more than 60% in the past 40 years. In this case, the decline may be equally due to loss of breeding and wintering habitat.

A few times on this day, I heard a Thicket Antpitta, so I selected its call with my player and speaker. The birds had been distant, and I didn't try very hard to call one. I'd become discouraged and was still ruminating on my excuses for failure. But now the call explodes from the trailside only a few feet away.

I quickly fumble out my player and hit "play." I have forgotten that I stopped the call in the middle the last time I'd used it. Also, I hadn't checked the volume on the speaker, which in my pocket has been turned to maximum. Consequently, what comes out is a truncated portion of a Thicket Antpitta's song played at a volume that would have frightened its cousins in Panama. It probably knocked down my own hearing a couple of decibels as well. The response, predictably, is silence.

Disconsolate, I sit on a log with my head in hands. Here I have a Thicket Antpitta close enough to call in so I could see it, and I have blown it. I know it couldn't have been enticed by the partial call I had just blasted into the quiet afternoon. *Ok,* I think, *Besides the other reasons I have for not seeing an antpitta, I need to add my personal incompetence.*

Eventually, I stop the self-deprecation and ask, *What if it happens again? How do I prevent this?* I have just demonstrated my tendency to become hysterical and make mistakes. I pull out my notebook and write a protocol. If I hear a Thicket Antpitta close by, I will first, try to decide the call's direction. Second, I will conceal myself as best as possible. Third, I will reset my player to ensure it will play from the beginning. Finally, I will start with a low volume so I can check the speaker before playing. I sit and memorize the protocol, telling myself I have a "four-step program" for nearby antpittas.

I look at my watch. It is about a quarter to four. The first hints of twilight begin about five—when I told Giancarlo I would return. I decide to sit and calm down a few more minutes and review what a good day I've had so I won't only focus on the botched encounter with the Antpitta. Suddenly, it calls AGAIN! And it is still nearby.

Now I know what to do. The call is behind me. I slowly scan the area to ensure there are no birds in sight. I stand and move slowly to a small tree to break my outline. I check the player and the speaker. And I wait. The bird calls again not far from where I heard it the first time. It is inside some very thick vegetation at a distance of at least 6 meters. I play my call. The antpitta answers, and the duel is on.

At first the contest is fun. I call, then a bit later, the antpitta answers. Eventually, I understand that it only calls while stationary. If it is silent, it is moving. Knowing this enables me to change positions a few times to improve my concealment and viewing position.

Again, and again, the antpitta calls. I answer. It is both exciting and frustrating to be hearing it right in front of me, expecting any moment for it to pop into view, and then have the next call come from behind after it has crossed the trail either on my left or right. Even though the trail was reasonably wide, the Antpitta circled me twice, crossing the trail four times without my seeing it. I begin to question whether the bird will ever permit me to see it.

After approximately forty minutes, as if becoming discouraged itself, it hops out in the open about two meters in front of me, and on to a fallen branch such that its perch was about ten centimeters off the ground. It points its bill slightly skyward, stretches out its neck and then lets loose its full-throated series of whistles. The best way to mimic the call is to whistle yourself, but by inhaling such that the whistle is a bit higher pitched. Do it five or six times rapidly, twice in succession, and you have a close approximation of the call.

The bird is squat when quiet, but when calling, I recognize that "full-throated" is an inadequate description. The neck is outstretched, but it isn't only the throat that quivers with each whistle, rather the chest and belly as well. I think of one of those old cartoons with a very fat Santa Claus who shakes when he laughs "like a bowl full of jelly."

I am so thrilled that I forget about my camera, which I let hang at my side. Now, I only want this antpitta to think it has won the long territorial duel in which it has participated, so I don't play my call again. Soon enough it hops back into the underbrush. I can still hear the call as I back away, flushed and delighted with the passion of my success. This is better than watching my team win the Final Four.

POSTSCRIPT: The new "science" of seeing antpittas.

In this chapter, I described my wife's sighting of a Scaled Antpitta, a bird I didn't see until a few years later. Now, my non-bird watching wife has seen eight species of antpitta. As for me, my number is up to sixteen. How did this happen? That's why there's this post-script.

I relate the first part of this story as I heard it from the world traveler and tour leader Victor Emmanuel, who was a speaker at a convention I attended. Ecuador, as a means of enhancing rural economies, provided some funding and training for some campesinos to develop eco-tourism. The focus for most was the Cock-of-the-Rock, a big, showy iconic species with grandiose mating rituals. They are a sight to see, and any natural-history traveler going to the countries in which they live will try to see them. Thus, farmers who had these birds on their properties were encouraged to build trails and blinds so they could share in tourist dollars and be less likely to log their property.

Because of that program, I wouldn't be surprised if more North Americans recognize the name of Angel Paz than any other Ecuadorian. Angel had a Cock-of-the-Rock lek on his farm near Mindo, so he built a trail and blind for their viewing. The trail work consisted of copious rock and dirt movement which he, of course, did by hand. As he dug, he uncovered worms and other invertebrates that were noticed by the resident antpittas. A couple began to follow him. Perhaps he noticed them and tossed a few worms their way.

While his work was in progress, a few birders were directed his way to see the Cock-of-the-Rock lek. While walking down the trail, a couple of different species of antpittas appeared, expecting Angel to resume his digging. A Cock-of-the-Rock lek is quite a sight, but there are a good number known and accessible. It is possible for anyone to see them, not so with antpittas. The birders went berserk and told their friends, who began to show up asking to see the antpittas.

Angel recognized the opportunity and began to collect worms every day and go out and feed the antpittas. He even named them and habituated some of them to his voice. "María," he would call, and out would step one of the desired species. Another was named José, and so on. Birders literally flocked to Angel's farm. Eventually, I was one of them, accompanied by my wife.

Angel has been very enterprising. When I was there, visitors were required to purchase breakfast (traditional and delicious) as part of the tour of the farm. A few tourist cabins were under construction. This enterprise has not only kept Angel from de-foresting his own land, but he has bought additional forest property and is letting some other land return to forest. His son was attending the university, the first youth from the area to do so—all because of the antpittas. Angel, being ambitious, has learned that some antthrushes are equally sought after, and I saw a Rufous-breasted Antthrush that came to be fed.

The birds are still wild and skittish. Angel was very direct in his admonitions to our group to be quiet and still while the birds were fed. Indeed, the most famous of his antpittas, a Giant Antpitta, did not show on the day we visited.

Our guide told us that after we left, Angel would go and spend whatever time it took to find the Giant Antpitta and feed it. Tourists do not visit year-around, but Angel and his family must feed the antpittas. It takes a great deal of time and effort to keep this enterprise going.

Word of Angel's success has spread. On that trip in Ecuador, and a later trip to Colombia, nearly every lodge had one or more antpittas habituated to feeding. That's how a non-birder, such as my wife, has seen eight species of antpitta.

Jocotoco Antpitta at a Feeding Station in Southern Ecuador

One of my guides in Ecuador and Colombia, an enterprising young biologist and organic farmer named Alejandro, was nonplussed. He lives near Angel and had conducted some research on his farm. Through Angel's observations and those of ornithologists who were attracted, quite a few facts have been learned and some previously unknown nests described. Not surprisingly, Angel will not let the scientists trap and band his antpittas, so no one knows how many generations are involved.

Alejandro's mixed feelings were like mine. Is it fair for "others" to see antpittas so easily, when I had worked so hard and felt so proud to find and attract one on my own? Shouldn't some sightings be relegated only to those who "deserve" them? Alejandro spoke of how these shy, secretive birds were no longer what "they should be." Casual tourists, and even birders who visited these lodges, were missing out on learning about antpitta's natural behavior. This wasn't even birding. It was like going to the zoo at feeding time.

Although he never expressed it, I also suspect Alejandro saw antpitta-feeding as cheapening his own skills and livelihood. As I've recounted, not all guides try to show their clients antpittas. Alejandro, being a biologist as well as a guide, was an expert at finding these birds and had led trips where the primary focus was to find antpittas. Now, with so many lodges feeding them, some of his special expertise is no longer in demand.

Then there is the other view. One well-known lodge in Costa Rica is in a formerly deforested area that was replanted with fruit trees and flowering plants designed to attract hummingbirds and tanagers. It isn't a natural forest but is a wonderful place to see many colorful tropical birds. Even if the deep forest species are missing, the new forest is better for wildlife than the previous cow pasture. Are the ubiquitous hummingbird feeders and fruit trays used at so many lodges any different than targeted plantings? If anyone can see an antpitta, will this enhance or harm the conservation efforts needed to save the increasingly remnant forests where they abound? And what about all of my effort to see a Thicket Antpitta? I know there are many who would cringe and be critical of my use of a call and my insertion in this bird's behavior for nearly an hour. Did I cause it undue stress and expose it to greater risk of predation for my own edification? Wouldn't five minutes of feeding per day be better for it—especially as an alternative to birders showing up every day to try and call it into view? These are all tough questions.

Nonetheless, when you consider what feeding antpittas has done for Angel Paz and his local community, the value cannot be denied. Clearly, the more humans there are, the more complicated becomes our relationships with wildlife. For my part, I am going to hope that the more people who see an antpitta, the more people there will be who want to protect more forest.

Despite the publicity regarding Angel Paz and those who have copied his methods in Ecuador and Colombia, so far as I know, no one has done anything similar in Costa Rica. If you want to see a Thicket and/or Black-crowned Antpitta, you are required to do it the old-fashioned way.

CHAPTER 6

BLESSED BY LIGHT

> *If people destroy something replaceable made by mankind, they are called vandals; if they destroy something irreplaceable made by God, they are called developers.*
> — *Joseph Wood Krutch*, The Desert Year

Ivory white, not creamy. White light! Unusual in the rainforest, but not uncommon today. I speak mostly of birds, but this is an orchid (*Sobralia macrophylla*). The blooms are at their peak for only a single day. The plant, with its six-inch flower, is gorgeous, especially when struck by a rare shaft of unobstructed sunlight.

Here's a bird, a tiny one. A Crowned Woodnymph. Barely four inches from head-to-tail, but with BIG colors. Bright purple on the crown, upper back and belly with an emerald green throat, the bird appears blackish when the light is dim, but when sunlit and iridescent? Oh my, a woodnymph indeed!

This is how my day started. Good start or bad? How to decide? Tropical birds do not like bright sun. Clear, hot days are the bane of tropical birders. Once, under a fruiting cecropia, I watched several colorful tanagers feeding actively. Suddenly, a passing cloud moved on and the tree was fully in the sun. All movement stopped. It was as if the birds had been abruptly vacuumed away. I peered intently with my binoculars and finally found a few of them motionless under the leaves. Their vivid colors became camouflage in the shade of the brightly illuminated leaves. I stayed and watched. Another cloud passed in front of the sun, and the birds became instantly active. I wondered how many times on a bright day I have passed trees full of birds and had no clue how many were so close. The best birding may be when there is incipient rain or mist. Birds are active then, but the colors are muted.

Hummingbirds, on the other hand, are made for light. Once, I was about to cross a stream on a Rara Avis trail on a bright sunny day. Sunlight was almost directly-overhead. I can still see the stream bottom glittering coppery-red in the bright sun. Dipping in the sparkling water was a flash of green, flecked with gold--a Black-bellied Hummingbird. I've seen them many times, but only once like this. The sunlight brought out its full iridescence, including the glint of bright rufous on its wings. The white feathers in the tail flashed strobe-like as the little bird bathed and splashed.

My second memory of that sighting was when I excitedly described it to the resident guide at that time. "Nope," she said. "You didn't see it. It's not on the list." I tend to be reticent and frequently too careful about sightings I know are rare. In this case, I hadn't known the bird had never been recorded at Rara Avis. At first, I was down, but I recovered. "No," I said. "I am not wrong on this one. There was too much light." And, a few days later, near where I had seen it, another birder reported the same species. I doubt he saw it like I did.

I enjoy my hummingbird memory, but today I am looking for other species, and the sunlight is not my friend. The British psychotherapist and essayist Adam Phillips has noted that "ambivalence is the basic condition of our lives." That is an apt description of what is occurring in my mind. The sunlight is a curse and a virtue. I am acutely aware of beauty I am certain to see if this brilliance continues during the long hike I have planned. I also know the brilliance and accompanying heat will cause most of the birds I seek to take quiet shelter.

On the other hand, it is the sun that powers not just this rainforest paradise, but our planet. I think of the energy of sunlight and how we use it. My love for nature and our planet tells me we've gone very far wrong with our use of the sun, and maybe it is already too late for adequate correction.

An oft-quoted statistic goes like this: more power from the sun hits the Earth in a single hour than humanity uses in an entire year. Just 200,000 square miles of solar panels with 20% efficiency would provide enough energy to power the entire earth in 2030. Unfortunately, humans choose to forego use of the ever-present sun and instead use millions of years of the sun's ancient energy concentrated in appropriately named fossil fuels such as coal, oil and natural gas.

By their nature, these fuels will run out, and at the rate humans are using them, they may already have damaged our planet irreparably. At one time, Costa Rica considered itself unlucky, because of the apparent lack of fossil fuels. Gasoline prices

(2018) are more than double the United States both because of higher taxes and because of the difficulty with which petroleum products are obtained. Fortunately, Costa Rica does not rely on fossil fuels for power generation, fulfilling their needs from wind, solar, geothermal and especially hydropower. Costa Rica has forbidden drilling and fossil fuel extraction, although the task was probably made easy by there being little with respect to reserves (www.ticotimes.net/2014/07/28/costa-rica-extends-ban-on-petroleum-extraction).

Unfortunately, no method of energy production is benign. When driving to Costa Rica's Arenal volcano area, I don't enjoy the wind turbines dotting the landscape, but I recognize how much better they are than a coal-fired power plant. (Limited studies have not shown bat or bird mortality associated with this installation.) Hydropower is much worse than wind because of the massive loss of habitat caused by dams and reservoirs and their combined effects downstream. Hydropower is at least carbon neutral.

Having banned oil and gas exploration, Costa Rica has spared itself the external control exerted by large private oil companies that distort the economies and governments of countries such as the United States. Also avoided are the always-present massive pollution and social problems. I saw these issues first-hand on a trip up the Río Napo in Amazonian Ecuador. On the small motor launch that we were riding to a remote jungle lodge, there was an Englishwoman who had married into the native community. While speaking with her, our boat was rocked by a large barge transporting an oil company tanker truck.

"These ships are a problem," she said. "They are huge. They travel fast. A few months ago, one swamped a family in a dugout. The entire family drowned. The ship didn't even stop. And, there's so much more bank erosion." She said erosion of a meter or more of bank was occurring every few months as opposed to what formerly occurred over several years' time.

Having worked with watershed restoration/analysis in the US, I had already examined the banks with a critical eye. It was plain to see what was happening. The river was at historically low levels. I have photos of a location where depth was marked on a dock's support piling. The landing platform was several meters overhead. The river was more than a meter below the lowest stripe on the piling used to evaluate depth of water. Indeed, our trip to the lodge was taking much longer than usual. The boatman had to slow several times to find the channel. We still scraped bottom on multiple occasions.

Two things make a river unusually shallow: unusual drought and a channel widened by misuse. I haven't seen data on Ecuador, but recent data in nearby Peru indicates more than a quarter of their glaciers have disappeared in the past few decades—to the extent that there is now an experiment whereby white paint is applied to rocks that previously were covered by perennial snow, in an attempt to mimic the microclimate by reflecting sunlight as the snow once did.

I was already familiar with the controversies regarding oil companies and eastern Ecuador. I had viewed the documentary *Crude*, and I had read a book called *Savages*, by Joe Kane, both of which recount the battle of native groups, aided by some NGOs, against oil impacts. The controversy was interesting on three levels. First, the natives who were impacted knew they were owed something for all the harm they had endured. Unfortunately, they were not sure what they should try to gain from the situation, leading to inconsistency and selfish actions on the part of some individuals. Second, the dedication of the foreigners who tried to help—virtually all women and referred to, frequently with derision, as "ecochicas," was very impressive. (The Englishwoman told me that if she attended a meeting with oil company officials, a "white face" in the room meant the meeting was over. The oil representatives refused to meet.) Third, I was impressed with the ease with which the locals could demonstrate massive contamination at supposedly cleaned-up locations, and how that contrasted with the naïve sincerity of the environmental officials with the US companies, some of whom appeared desperate to believe they had done a great job restoring the same sites.

I have thirty-some years of personal experience with field environmental conflicts and cleanup activities. It was all familiar. Despite what the attorneys for the natives said, the oil companies probably had the paperwork to show they had followed the letter of the law. These documents were the grounds for the impassioned rhetoric by US environmental officers as they defended their companies. These folks are paid very well. They do not want to endanger their standard of living. They want to believe the best occurred.

I know what probably happened. Even in the US, there is a double standard for oil and gas companies. Several times I heard the owner of an environmental consulting company I've worked for explain how a task was to be done and say, "[the good way] is how it's done. Unless, of course, you are in the oil patch."

More recently, with the oil and gas drilling boom in Colorado, a friend of mine with another consulting company told me one of his clients insisted he sample streams only after heavy rain or during spring run-off. The hope was that doing so would ensure contamination, which might show up in base flow, would be diluted.

Letters to the editors of local papers from former workers have described orders to bury wildlife that died from inadequately protected waste pits. At least two illegal operations have been found in our county and one that is adjacent. What did county officials do? "Oh, the poor company has so many regulations to deal with. It was an oversight." And, they quickly voted to permit the operations.

Oil and gas extraction have had such a different ethic for so long, that I doubt it is even visible to most of the people in the industry. My own work in oil and gas was limited, but I saw enough. I have been lied to on several occasions by high-ranking environmental personnel. In one case, officials representing one of the largest utilities in the world insisted so passionately that our government-mandated "check" of their site was such a waste of taxpayer money that I and my staff were apologizing for our presence—until our cursory drilling program hit illegally-buried waste drums. The last I knew, cleanup of the site was well into the millions. The utility company was not prosecuted because they blamed inadvertent oversights, mistakes in record-keeping, etc. Besides, they were the major employer in the region.

Another time, while working for a well-known Dow-Jones Industrial company with a large fossil fuel division, I listened to a company-wide lecture from the head of environmental affairs exhorting his troops to be leaders in the industry in terms of environmental compliance. Later, in a smaller meeting, encompassing only the heads of the major company sectors, that is, those in charge of ensuring the many factories were profitable, this same man said, "Now, you know I don't give two s$#&s and a holler what goes down the river if a plant is losing money."

Sadly, this behavior isn't limited to the polluters. I was a participant in a meeting discussing an extensive environmental site-characterization that was to be done at a large industrial plant. The project was to be performed by a multi-national engineering firm. The technical personnel unveiled their plan, which matched the budget they had submitted for winning the project. They were quickly over-ruled by a vice-president/program manager. "Add 20% more of everything [meaning sampling locations/boreholes]," he said. "They'll go for it." One of the other technical people leaned over to me and said, "He gets 10% of any contract change-order."

You can see how it is difficult for me to live in a time that so idolizes business and the "invisible hand" of the market. I too believe in those principles, but I also recognize that too many humans aren't to be trusted, especially when it is possible to conceal bad deeds within the mass of a large corporate entity. Hence, we need checks and balances, which means regulations, and regulators, and a legal system that reliably provides justice.

I've also known of short-cuts taken by cleanup contractors. *Bury it. Don't remove it as you promised. Bribe the inspector. After all, this is a money-making business.* You obtain the work by low bid. If something goes wrong or takes longer, what do you do? Many find short-cuts. So, what happened in Ecuador? Well, what manner of companies and workers would take environmental cleanup jobs in sub-tropical Amazonia? Much of this cleanup activity occurred deep in the jungle far from connecting roads where heavy equipment had to be shipped in on the river. How good were inspections by local officials who may have been inhibited by inadequate training/experience or vulnerable because of low-pay or even government "encouragement" not to discourage the business?

One last oil anecdote! I once shared a long plane ride with a contractor putting in pipelines for a major oil company in Venezuela. He had a few too many drinks on the flight and began to tell me some of the environmental atrocities his company was committing—large oil spills and other leaks of contaminants, as well as damaging dredging of rivers, mangroves and seashores. "I know I should be thinking about my kid's futures," he said. "Then I think, screw my kids. We can make this money now and we've got to do it."

This is how it works. He didn't work for the oil company, per se. He was a contractor. His reports went to the oil company who received them, perhaps, in good faith. The oil company may have known or suspected the damage, but they had the contractor's reports to hide behind. If problems were ever discovered, it wasn't their fault. Besides, how do you deal with a problem so far away? It isn't like there aren't extenuating circumstances, other people's jobs, stockholders etc. Now, the offending subcontracting company probably no longer exists, having taken the profits, disbanded, and reformed as a new entity. That was the final take-away that I gleaned from *Savages* and *Crude*. Everyone agreed, the environmental problems were severe. They were caused by oil extraction. The tragedy was that it was no one's fault.

Finally, to those who would say (as has right-wing commentator Rush Limbaugh) that environmentalists should be blamed for forcing oil production to locations where it is more difficult to extract without impact: I have a book written by my hero Alexander Skutch about his 1920's visit (he was looking for rubber plants) to Eastern Ecuador—this very area. He made note of the oil exploration occurring even then.

So back to my conversation with the Englishwoman who had married into the indigenous group. She shuddered when I asked her about the oil company impacts. Her people had observed what happened to other nearby indigenous groups and had written the government and said essentially, "So long as one of our people has breath, there will be no drilling on our land." The oil companies, however, had a desperate need for a pipeline through their land. There was so much pressure from the companies and the government that a deal was finally struck. "Build us a lodge," said the natives. The people, again having seen what happened with other tribes and their dealings with the oil companies, insisted the lodge be completed before permission for the pipeline was granted. Otherwise, they believed, the company would renege on its promise. That is how our destination, came to be—complete with a massive tower that helped make birding so great. The lodge provides jobs and some money for the tribe, but, my contact said, they are still impoverished.

"There are a lot of impacts," she said. "Our only water source is the river. We bathe in it. Drink it." Indeed, we often saw people bathing or washing clothes, and yet the barge and ship traffic was heavy. "The river is polluted now. Gastro-intestinal problems and skin rashes are a much greater problem than anyone can remember. "

I asked her what she does about drinking water. Her response: "I have money, I only drink bottled water. The one time I ran out, I boiled river water, filtered it, and put tablets in it."

Most US citizens do not recognize what a reality climate change is to the rest of the world. Our young Peruvian guide, a few years ago, remarked to me while waving at the beautiful Madre de Dios river and what seemed an unlimited river (*zarzea*) forest, "This will all be gone when I'm old anyway."

On this Ecuador trip, the Río Napo was wide and slow because of the unusual drought and heat everyone was complaining about. It was also being widened by the river traffic from the large oil company barges and ships. At low river levels, boats are confined to the deepest part of the channel, known as the *thalweg*. Simple physics regarding dissipation of energy requires the *thalweg* to take a sinuous course which will bring it near the banks on outside turns. With small native dugouts, those banks aren't affected, but large ships and barges send waves against the banks, causing the erosion known as mass wasting, that is, large slabs of bank falling in. Thus, as snowmelt lessens and traffic increases, the river becomes wider and shallower.

Wider and shallower increases the water temperature and changes the channel substrate. Typically, more fine sediment accumulates, which decreases the productivity of the small plants and animals that are the basis of the river ecosystem. The Englishwoman said her people were complaining more and more that the fishing was not what they were accustomed to and was deteriorating year-by-year.

The oil business has given Ecuador good roads, but it has taken much away from their present and our planet's future as the damage continues. The people I was visiting live in a biological fantasy land under siege by fossil fuel extraction.

The usual responses I receive to my complaints about the damage from fossil fuels are "jobs" and "economy." I am not sympathetic. Oil company profits are enormous—most years with double digit returns on investments. What if that return was only 8 or 9%? That would still be a good investment. The rest of the money could be used to provide more jobs—jobs for limiting the destruction during extraction and for ensuring adequate restoration afterwards. I am not so foolish as to underestimate the amount of energy our world needs; there will be environmental losses, but they can be limited, and in many cases could be rendered temporary.

I'm put in mind of a surface coal mine I know of in Northwestern Colorado. The restoration manager showed me some reclaimed area. I provide the caveat that I have no idea of the overall balance of extraction versus restoration on this property, but the restored area, now kept free of grazing and planted with native shrubs and grasses, was a showcase. If fossil fuel companies used that sort of ethic for all their site restorations, I could become a cheerleader for them—except, of course, for climate change.

Costa Rica, with its remarkably high biodiversity for its small size, illustrates the world-wide effects of fossil-fuel-induced climate change in microcosm. When my family visited the first time in 1989, the most famous natural location in the country was the Monteverde Cloud Forest. Although I now know better, at the time I thought this was the only place to see a Resplendent Quetzal—at least a candidate, if not a winner, for the title of world's most beautiful bird.

Alas, our visit was beset by high winds and copious rain. Quetzals were not to be found. Although there were birds to see, I didn't realize how inadequate my desert binoculars would be in the rainforest. They quickly fogged up, presenting me with only smudged views of such exotics as Black Guan and Red-headed Barbet. Alan, our guide, was elated. It had been dry. He told us this rain was called the "Caribbean Rain," because it came from the east and was perfect for the iconic Golden Toad. I have a 30+-year-old brochure from that trip—prominently displayed and advertised is the Golden Toad.

Alan explained that Golden Toads breed in the top of the forest along the Continental Divide. Golden Toads, he said, spend most of their lives underground emerging to breed only after appropriate rains. A few weeks later, the adults resume their subterranean existence, but they leave the remaining pools of rainwater full of tadpoles. Alan was excited because he was part of a group that was going to spend several successive nights doing a census of this beautiful toad. He related that the appropriate rainfall had simply not occurred the previous year, and no toads were found. Everyone was optimistic this year.

I learned subsequently, that after several days and then weeks of increasingly frantic searching, no toads were found—not that year, nor the next, nor in any succeeding year. The Golden Toad is now classified as extinct. Many studies have searched for the reasons for the Golden Toad's demise. Doubtless, a fungus delivered the final blow, but nearly every examination of the issue has pointed to climate change (drought and unusual temperatures) as a contributory and necessary link in the Golden Toad's demise.

In most of the US, climate is sufficiently variable that almost anything can and does happen on any particular day. That is a tragedy for the world, because "crazy weather," normal for the US, has given cover to climate deniers. A changed climate is apparent when there are distinct wet and dry seasons as in Costa Rica. Consider that a primary reason for Costa Rica's remarkable biodiversity is the existence of so many micro-habitat niches, each inhabited by species for which that niche is just right. When the system is so delicately balanced, there is no such thing as a minor change. When climate changes too rapidly as it is now doing, there is no time to adapt. One of my bird-guiding friends said this in a recent blog: "Something is happening.... we don't have ...an immediate way to lower temperatures and increase the rains. The effects of higher temperatures and drought in forests that are adapted to four meters of rain...seem to be showing...We see and hear fewer birds in those forests...Dawn choruses were very quiet in places that experienced truly fantastic morning song at the same time three and four years ago. ... If some of the most intact rainforests in Costa Rica are like this, I can't help but wonder how many areas are approaching ecosystem collapse. It's not just a drought, it's prolonged hot, dry weather caused by global warming in places not adapted to those conditions, and the outlook is bad" (Pat O'Donnell, birdingcraft.com/wordpress/2016/04/28/).

A scientist quoted in a Costa Rican English language newspaper, put it this way: "There have been changes in the temperature and rainfall. There has been a decrease in the amount of drizzle and fog that comes with the trade winds of the Caribbean, which are really important, especially during the dry season. We have perceived a change in the variability of rainfall; it's not that it rains less, but rather it is more variable, causing longer dry periods that didn't exist before. In the long term, there are more days without rainfall" (Alan Pounds, Jan 9, 2016 *Tico Times*).

My own observations bear this out. When I initially visited Rara Avis, I was amazed at the abundance of tanagers, not just the diversity, but the sheer numbers. We visited several times in the early 90s and by the time we returned after a decade's hiatus, it wasn't the same. The birds were still present, but less common. The Cornell birders I had met in 1991 had referred to many of them as "junk tanagers" because they were so abundant. They only became excited with the sighting of the rarer Black and Yellow Tanager and especially with the Blue and Gold Tanager, which only lives in a narrow elevation band. I recall how much more often I, as a rookie tropical birder, encountered large mixed flocks, but not so much this time.

On previous trips I had also focused on observing Heliconias. One, with a popular name of "lobster-claw," was abundant. The long tubular flowers are beloved by hermit hummingbirds, and I watched them especially to see the strange-looking Long-billed Hermit. The Long-billed Hermit's body is about the size and color of a small, wooden clothespin, but then out the top is a long-curved bill, and out the bottom is a long, thin tail tipped with white. It is bizarre-looking and a favorite of mine. Watching Heliconias also leads to the possibility of seeing the much-rarer bird, Rara Avis's emblem, the White-tipped Sicklebill.

Sicklebills feed mostly on Heliconias because, as their name implies, they have a long-curved bill that is perfect for feeding on the tubular flowers. The only problem was that this time, as I had noted on a previous visit, Heliconias had become rare. I could hardly find any. I don't know if this is a case of forest succession or climate, but even to my unskilled eyes, there had been a lot of changes since my initial visits.

I cannot help but think of all this on my walk. It is the beginning of the rainy season, but this day is bright and sunny. Sure, days like this have always occurred during the rainy season, but are there more now than usual? Is all this light why I am seeing so many changes?

I have planned a long hike. I have no idea of the origin of the names for some Rara Avis trails, but I like the sound of Margot to Guacimo to Levi to Atajo (shortcut) to Catarata. Well, the last one at least is obvious; the trail passes the double waterfalls near the lodge.

In keeping with the bright sun, I do not see much activity in the trees, but the wrens are loudly at it. Most everyone is familiar with wrens, perky brown little birds usually easy to hear but difficult to focus on because of their busyness.

Wrens are reasonably represented in North America. In the Midwest, where I grew up, six species are theoretically possible, but only one, the House Wren, is commonly seen. As a youth, I loved seeing them in the yard. I can still see one in my mind's eye in full-throated chortle as it sang its beautiful little song from my mother's clothesline. In the Western US, it would not be too difficult to find five species in a single summer day. In the mountains, once again, the House Wren is ubiquitous. It is also very common in Costa Rica, but it is a yard and garden bird, a different subspecies, once considered a full species on its own. Here at Rara Avis, eleven species of wren have been detected and I quickly encounter three of them.

As I cross the river, I hear a Bay Wren. Bay Wrens are quite a bit more handsome than the mostly tan House Wren. I hear him first. The song is a loud whistle followed by a "whee whit." Mostly rufous with a black cap and bold white on the face, one announced its presence almost every time I crossed the bridge. I liked thinking it was saying "hello," but it was more likely "beat it!"

Down the muddy road I walk and a loud "freee-oo-whee, freee-oo-whee" announces a Stripe-breasted wren. He is also handsome: boldly striped white and blackish-brown from belly to throat, with a black line through the eye and black barring on the wings and tail.

Soon I arrive at the Margot trail and as I ascend, I hear another of my favorite sounds of the rainforest, a loud high-pitched but musical "whee-whee-wheeup" repeated over and over. Sometimes the little bird adds a brightly whistled three-note call "whee-whee-whee," which in relative pitch and cadence makes me think of "three blind mice" sung at a high pitch. The call, as beautiful as it is, startles me. I was lost in thought. The forest was quiet and then suddenly the loud song burst from the trailside. I'm listening to a White-breasted Wood-wren—frequently heard, but seldom seen. I know he has a creamy white breast, short tail, boldly patterned brown and white face with a dark line through the eye. I could probably see him if I wanted to play his call or spend some time, but there's no need. I'm just happy to hear him and glad to be on a trail in the rainforest.

This is the life, I'm thinking. *Or death!*

I am one step from a fer-de-lance! Curled up right in the middle of the trail, I spy black-edged diamonds bordered in creamy yellow, and a triangular head. This is one snake you must be able to recognize—and avoid. They are common here. This was one of two I saw on this trip.

Rara Avis has always been a place where it is possible to see many snakes. When my wife and I were here two years ago, an Eyelash Viper spent several days in a shrub right next to the table where we ate. It was a peculiar feeling to pass within a few feet of a deadly serpent several times a day.

Usually, at Rara Avis, I encounter one or more hog-nosed pit vipers, relatives of rattlesnakes. Unlike the rattlers, these don't have rattles, but they are recognizable by the same triangular head. Fortunately, I've never encountered any that were aggressive. They lie curled up by the trail and pay no attention to passers-by. One guidebook noted one of these was observed to stay in the same location for 19 consecutive days.

In this instance, it would be inaccurate to say I had a close call with this fer-de-lance. I was watching the trail and saw the snake before taking the step that could have had serious consequences. The Florida State researchers were working in this area and came up the trail not long afterwards. When I returned that night, Jason ran up to me and exclaimed, "was that your footprint so close to the fer-de-lance?" I guess it was, but I saw it and backed away easily enough. Still, it worries me. I do become engrossed when looking for or at a bird. How sure can I be that I will never miss a fer-de-lance? Or a bushmaster, also found in this forest?

Not surprisingly, the warmth and sun make it a good day for snakes. They enjoy what sunny patches they can find. I see snakes the biologists later help me identify, such as a Lower-montane Green Racer. I love to see and identify wildlife, even snakes, but I am perfectly happy not to find the venomous kind.

I finally reach the Levi trail, a several-mile route on the edge of the property through an area that is naturally drier by being near a steep sun-facing slope. I am not enough of a forest ecologist to be sure, but it also appears to be second growth rather than primary forest. The aspect is different than the other forest where I have been hiking. Most of the trees are smaller, and the understory not nearly so thick.

The sunlight dapples through the trees. The combination of sunlight and a lower canopy seems to cast more brown and yellow in contrast to the deep green I've grown accustomed to. The forest floor is a tan and sun-yellow plaid. Perhaps I notice the color because nothing is happening. I frequently hear lizards scurry off into the brush—the sudden rattle of leaves giving my startle-reflex lots of practice, but I don't see or hear much else.

As usual, I have a bird in mind—a Song Wren. This unusual bird looks more like one of the antbirds, having blue about its eyes and colors of blackish-chestnut. This bird is a problem for taxonomists, there being only three closely-related species in its genus. It isn't quite a wren and it isn't quite an antbird, just another of the many marvels of evolution in this amazing world. It also isn't very songful to my ear. The call is a combination of whistles, squeaks and what I can only call high-pitched growls. I have never seen one. One of the few calls I hear all day is a Song Wren alongside the trail, but it is out-of-sight, and stays there despite my attempts to entice it closer by playing the call.

Now, at least, I am quite aware of nature's creative energy as the battle for scarce sunlight gives rise to some interesting adaptations. A relatively common plant at Rara Avis is the Walking Palm—so-named because it has aboveground roots which can move the plant toward light. It can't move more than a couple of feet, but that can be enough to give it sufficient edge to survive amidst all the competition.

The struggle for light is the reason light gaps are so important to rainforest health. In a natural setting, these gaps occur when a tree falls and brings down most of the surrounding vegetation. Pioneer trees, such as cecropia, need a lot of light and are the first to sprout and grow. Over time, the fast-growing, sun-worshipping trees are overtaken by slower-growing giants which will, once again, complete the canopy overhead.

Light gaps provide some extra diversity and are excellent for wildlife. For one thing, it is possible to see a little farther. Here, as I walk, I muse that much of the trail is overgrown. Often, I detect birds passing just a little too far in front, or a little too high overhead.

The Levi trail ends, and I'm back at the tractor road. In the old days, it was called the corduroy road because it was a series of logs laid parallel to each other. Many, many logs have fed this road and mud, and I wonder how deep the logs now are. In several places, the road itself has sunk or been excavated up to five or six feet by rotating tractor tires.

This is a damaging road, as are the trails. I've seen what happens during a rain. The uphill side of the trail shows clear water flowing down the hill. The downslope side of the trail is bloody red as the soil is washed away. That mud and silt is affecting the creatures that would or should live on the downhill side, causing unseen damage besides the soil loss.

The road does provide a light gap of sorts, so I decide it is a good place for lunch. I can see quite a bit farther here and, as I've often experienced, the best way to see something is to sit still and wait. No such luck today; my lunch is uneventful.

On a previous trip, I had several successive bright, dry days like this, and I chose impatience and frustration as my manner of dealing with it. Almost too late, I changed my focus to search for plants, insects, reptiles and amphibians. Then, the magic re-appeared. That's what I do today.

Along the road, I kick at some leaf piles and out hops a bright red and blue, little frog. These guys have several common names because the extent and location of the coloration varies, but many, such as this one, have blue hind legs while the rest of the body is carmine red. For that reason, they are called "blue-jeans frogs" or "strawberry frogs." Because of the bright colors, I know they are of the genus *oophaga,* or poison dart frogs. This one's scientific name is *Oophaga pumilio.*

It strikes me as interesting that fruits can be bright red as a way of saying "eat me," while this guy is brightly colored as a warning, "don't eat me." It all has to do with reproduction. Fruits want to be eaten so their seeds can be scattered, and poison frogs do not want to be eaten so they can go about their business of finding a mate and laying eggs for the next generation.

The reproductive strategy of these tiny frogs is very interesting, It begins with vent-to-vent copulation rather than amplexus—the usually-depicted method of male grasping female while clinging to her back. Three or four eggs are laid, and the male watches over them—often transporting water in its cloaca to evacuate on the eggs and prevent dehydration.

After ten days, the eggs hatch and the female transports the tadpoles on her back to a water-filled location such as a concave leaf, a bromeliad axil or a depression in the soil. Single tadpoles are deposited at each location. The female then comes to each tadpole and deposits unfertilized eggs as food. *O. pumilio* tadpoles are incapable of using any other form of nutrition. In a month, the tadpole metamorphoses into a small frog.

The road, I realize, is a highway for butterflies. I keep seeing one that is bright yellow with a black border. The researchers later identify it without hesitation--a Mimosa Yellow (*Pyrisitia nise*). I later read that it is found from the Southern US to Argentina. Such a mystery. Why do some species use a small area, while others, closely related, inhabit thousands of square miles? Birds are the same. Some have minuscule ranges, while others, seemingly very similar in habits, range widely.

Another common butterfly, which I erroneously think is in the *heliconius* genus, is Isabella's Longwing (*Eueides Isabella)*—yellow and black on the wingtips with orange and black stripes on the inner wings. Many of them flit back and forth across the trail as if many flowers are fluttering in the wind. Isabella's Longwing is a true tropical species living from Mexico to the Amazon Basin. Finally, I was dazzled by a butterfly with lots of blue--the blue-winged euyrbia (*Eurybia lycisca*), another tropical species found from Mexico to Ecuador.

My mind wanders. Again, I am forgetting Thoreau's admonition "to only think of the woods." My rumination at such times usually goes to the big questions. *Somewhere there is conflict. Somewhere there is starvation and illness. What good am I to be walking in the wilderness?* I think again of my mental mentor Skutch and his belief that only appreciation of nature can unite humankind. If so, then my appreciation and experience of it is a worthy use of time.

A question I ponder is whether there are enough of us to save places like this. Can we convince the rest of the world how worth saving they are? What would it take? As I write, the world is watching the Trump administration. Very important people who have power over areas such as this include his former Secretary of State who has presided over the world's largest oil company. However one combines or separates population growth and the fossil fuel industry, these are the two biggest factors in the loss of wildlands.

My rumination continues. *Light! Light! Focus on it,* I tell myself—which leads me to consider several small buildings I've seen since my arrival in Costa Rica. "La Luz del Mundo" proclaims their signs. The "Light of the World." My mind again strays to my Roman Catholic upbringing. Almost daily I went to mass and recall, at least in paraphrase, the famous gospel of John about the light shining in the darkness but the "darkness grasped it not." Metaphorically and stylistically beautiful, I tried hard to make that "light" work for me, but it didn't.

The light that works for me is the light of evidence. Evidence is what leads to truth. I deplore the massive amounts of money spent on buildings and ritual when those expenditures are not based on fact. Wouldn't we get more bang for our buck if we spent it on what we know to be true? I know it is ridiculous to hope for my type of conversion in the short term, but I remain convinced that science has most of the answers already. I know it is said that those of liberal or progressive thinking universally believe that if they could only sit down with their adversaries and share what they know, it would be convincing. This idea is generally mocked, but personally, when it comes to the issue of religion, I believe it.

No, I'll confess that I don't just believe it, I *know* it. I cringe when commentators dismiss any debate between science and religion by suggesting that scientists see their views as their religion. Not so. I will admit that the explanation may not be satisfying to a non-scientist. E.O. Wilson described it as a "certain way of knowing," that is, how a scientist can "know" the truth of work in which he has not participated. And, this "knowing" is not the same as someone who "knows" the truth of their chosen religion.

As with any endeavor, there are scientists who are frauds. Something "proven" by one person is not what I'm talking about. I can explain this best by considering the issue of prayer. I have experienced a deep, restorative, transcendental mental state from prayer. When a religious person talks of such an experience, I believe them. It is real. I can think of dark nights in a pew when I gave myself to experiencing the overall-encompassing love of a supreme being. It felt wonderful. I departed the church believing such a being existed. Being able to have such an experience was one of my proofs.

But, after a while, I began to ask how I could explain why this loving being let my mother suffer horribly and die young from cancer. Why did he set up rules that would prevent me from marrying the person I still believe to be the "best person in the world," because once she made an honest mistake and entered a disastrous marriage? Those are only two of my personal examples. How easy it is to keep the list going: wars, unexplainable accidents, natural disasters, childhood disease, being born in a bad place, and poverty. Add your own. All part of "The Plan," some say.

When my son's roommate died in a senseless accident, his mother-in-law to be, whose husband had died from an even more unlikely accident, greeted him with, "If there's a plan, it sucks." Then there is the story I heard on PBS's "This American Life," about the preacher in the mega-church who, after watching accounts of infants born into extreme deprivation on the news one night, realized, "The god we worship is a monster." He remained a Christian but dropped the idea of a personal god who could be manipulated by prayer or who had some sort of cogent plan. Unfortunately, when he began preaching a gospel consisting only of love and inclusion without hellfire and damnation, his large congregation quickly dropped him.

Which brings me back to the experiences I had when "deep in prayer." My desires to understand the human condition and how best to live have kept me seeking a functional philosophy of life. Ultimately, I duplicated that prayerful experience in meditation—secular Buddhism meditation. Again, that's just my experience. If it were only about me, I could say no more. However, scientists have developed sophisticated ways of examining the brain and have proven, simply by watching which parts of the brain are activated, that a deep prayerful experience and a deep meditative experience, sans god, are physically the same. That's the "light" of knowledge.

Secular Buddhists emphasize that their practice is not anti-religious. Anyone of any religious persuasion can use the meditation techniques because there is no dogma attached. What then is the need for dogma? That's what people usually fight over—or use as an excuse for fighting.

Much of this explains why atheism is on the rise. There have been prominent discussions regarding the efficacy of religion and existence of god before, but that was all philosophy a la Bertrand Russell. Now we have hard data from physical scientists. A recent issue of *Discover* magazine reported on five different methods by which god-like or spiritual experiences are being explained by physical methods such as genetics, neuroscience and biochemistry. I certainly wouldn't feel free to expound on this without seeing some of the scientific evidence and, in some instances, cultural change. Armed with this new information, we are obliged not to misuse it.

One of the saddest things I ever heard was when my mother-in-law, a true believer in her own mind in Protestant Christianity, said something to the effect that if the dogma she believed wasn't true, "What good was anything? Why do anything good?" I had to stop the conversation at that point or tears would have inevitably followed, but I wished I could have asked her a few questions such as, "If you were certain there was no afterlife, would you still love your kids? Would you start cheating on your income tax? Would you cease giving to any charities?" I know her answers would have been the same whether she believed in the dogma or not. She was a kind person at heart, but I could never understand the god she believed in. He would get all the credit for anything good that happened, while tragedy was "part of the plan that we poor humans couldn't understand." She wanted the solace of believing that the inevitable and normal suffering in her own life was part of a plan imposed on her by a loved one who somehow needed that suffering from her. How could she understand a world if suffering has no meaning?

Instead, suffering has meaning precisely because we are all part of the same family and we all go through it. Life's randomness parcels out suffering in a manner that is often inexplicable. That's why we need each other in community. That's why we shouldn't eliminate the rituals so precious to so many. Indeed, I agree with author/philosopher Alain de Boton and his "ritual for atheists." Doubtless, most "religious" people only "believe" because they treasure their participation and acceptance by the "community."

With my own eyes, my atheist self has marveled at the beauty of religious art. Same with my ears. Who cannot deny being stirred at the beauty and triumph of Handel's "Hallelujah Chorus?" Alain de Boton's "rituals for atheists," correctly recognizes that uniting over appreciation of beauty is wonderful and necessary. Note the echo of Skutch and his belief in uniting over an appreciation of nature. Sadly, mostly, man sullies the beauty by loading it with dogma. There's a lot of good science on this and how the parts of the brain and hormonal secretions that make us feel good promote our appreciation of beauty and need for community. So much could be addressed unambiguously by relying only on the light of evidence.

As I've grown older and continued to study philosophy, science and religion, I am attracted to Buddhism, not as religion, but as a "science of the mind," as one of the West's most important teachers, Jack Kornfield, has described it. Western Buddhism is a curious thing because it encompasses the deeply-religious and the secular. I have experienced a week-long silent retreat and was part of a *sangha* (sitting group) for several years. Aspiring Buddhists are no different from aspiring Christians in that I encountered some who used extensive study to try and understand and learn how to "do it right." That led to discussions of what was "real" Buddhism.

After my experiences with Christianity, those discussions left me cold. I've heard of the disparagement of "cafeteria Christians" who would "pick and choose" what was meaningful to them. That's what I found endearing about the Buddhist teachers I admire most. I mentioned Kornfield above, but there's also Jon Kabat-Zin, another of those who have pioneered the use of eastern meditation techniques in this country. His work has been extended to show many health benefits and even changes in the brains of practitioners. In Kabat-Zin's view, everyone can profit by embracing Buddhist principles, no matter whether they are Catholics, Jews, Muslims, or atheists. Indeed, I listen to dharma talks, the Buddhist equivalent to a sermon, several hours most weeks, and was surprised, and impressed, when I learned that one of my favorite teachers was a Catholic nun.

I know there are Christians who question what at first can seem a lack of "meaning" in western Buddhism. I don't agree. Perhaps superficially, but not ultimately. For me, the meaning of life in Buddhism comes from the four Bramaviharas (sublime attitudes or virtues). Live according to those and life won't lack meaning. Briefly, here they are:

- *Loving-kindness* —Everyone is on the same journey. As another Western teacher, Sylvia Borstein, says, "Life is so difficult, how can we be anything but kind?"

- *Compassion*—When you look at everyone you meet, hold them in your attention, and say to yourself, "This person wants to be happy. This person does not want to suffer." Maintain that attitude throughout a day and you will find plenty of meaning.

- *Sympathetic Joy*—Yes, be joyful, but focus it on the happiness of others. I see it as recognizing that we are all part of one community.

- *Equanimity*—Now we are back to Buddhism's Noble Truths that pain and death are part of life. Make room for their inevitability. Accept them. Learn what they teach. What could be more meaningful?

I should be a well-actualized, accomplished Buddhist, wouldn't you think? I compare it to my study of Spanish. Give me a written test. I will do well. Putting it into practice is a lot more difficult. I had such an opportunity following an accident that required multiple surgeries, a long convalescence, and continued management of the injuries through exercise and therapy. How did I do? How am I doing? Often not so great, but I can say unequivocally that I was and am better because of Buddhist practices.

During the worst times, I kept a small tract in my pocket written by another Western teacher, Pema Chodrin. The tract is entitled "Awakening Loving Kindness." A friend brought it to me while I was in the hospital. I had no idea she had an interest or exposure to Buddhism. I asked her about it. She was non-committal. "I picked it up somewhere," she said. "I thought maybe you would like it."

As I struggled with recovery for months thereafter, I carried a timer, set for an hour. When it rang, I would reset it for 5 minutes and read the little book. It always helped. I've said, realistically I believe, that "if my misery level was 11, as it so often seemed, on a scale of 1-10, 5 minutes with the book brought it down to a 9." (I know what a Buddhist teacher would say about that level of success: "Oh, very good. Now you can work with being a 9! What does that feel like?")

My mother's characteristic response to pain was "offer it up." Yes, if I was able to grin and bear it, some soul I didn't know escaped from purgatory a day or so early. In that way, feeling pain, a natural part of life, engendered guilt. If I was a wimp and didn't bear my pain, this poor soul burned a day longer. Thus, to me, if Catholicism had a first Noble Truth it was, "All life is suffering, and it is your fault!"

Experience and maturity have taught me that I was more serious and sensitive than most children. I still tend to take words literally and specifically. I know that Catholicism has recognized the damage it did in cases like mine, and many of the stories and admonitions I was subject to no longer occur. I know that the way it was for me didn't have to be that way. I know others were not susceptible to the guilt and low self-esteem as I was. The Catholic Church stands for many wonderful things and has done much to relieve suffering in the world. I have family members for which it is the center of their lives. I mean no offense. I can only describe how it felt to me.

Some of my feelings were probably related to my eventual choice to become a scientist. I'm far from the first to recognize that "faith" and the "scientific method" are natural enemies. I see nothing wrong with being a reductionist and believe we should learn and then apply all we can. For example, Richard J. Davidson, in his book, *The Emotional Life of Your Brain*, describes a study in which he divided his research subjects into two groups. One group was taught cognitive behavior strategies, which involved attempting to change beliefs, attitudes and ideas towards others. The second group was taught loving-kindness meditation. In this practice, the meditator mentally extends love and care to others. Both groups were instructed to practice 30 minutes per day over a period of two weeks. Brain scans were conducted before and after the two weeks. Following this modest amount of practice, the group that practiced loving-kindness meditation showed significantly more change in the area of the brain that predicted pro-social behavior. I have hopes that schools can use studies such as this so that everyone can improve their abilities to cooperate.

Once again, I realize I am wearing myself out by thinking too much and residing deep in my mind. *Time to get out of there!* I complete my sojourn on the tractor road as I reach the Atajo (shortcut) trail. I re-enter the dark jungle. The trees are larger than they were on Levi. The forest is darker and quieter. I continue my search for whatever I can find. Here in the gloomy, green, shady darkness, I find a large toad: *rhinella marina* or Cane Toad. He could weigh a pound or two. I've read they can weigh more than three pounds. I find litter geckos (*Lepidoblepharis xanthostigma*) a couple of times, hiding in the dead leaves. Litter skinks (*Sphenomorphus cherriei*) dash off through the leaves. Even better are the frogs. I spy a Brilliant Forest Frog (*Rana warszewitschii*). The name goes a bit far, I think. It has red-webbed feet and small green spots on its back and yellow spots on its thighs. I saw several others, even a couple that I photographed, but still haven't identified. I have two guidebooks to Costa Rican reptiles and amphibians, but they only scratch the proverbial surface of the possible species. I've tried on-line identification sites and I've still failed. Someday I'll figure them out. I'm amazed at how many I see because most are so well-camouflaged. Then dawns the realization that as difficult as it is to see these creatures, there are likely many more that I'm missing.

The quiet is suddenly pierced by the loud raucous calls of Carmiol's tanagers. Their call is described by Stiles and Skutch as screaming, high, "staccato, nasal, scratchy...grating or weezy...often prolonged." I'm excited because the guidebook goes on to say that these flocks are often "joined by ovenbirds, woodcreepers, and other birds." I brace myself for a great encounter with a mixed flock. But I don't even see the Carmiol's. Other than some flashes of movement, I only hear them on one side of the trail, then overhead, then the other side, then gone.

As I trudge back to my room, I feel some disappointment. New birds today? None. Interesting bird encounters today? Not so much. But then I think of the light: the light of iridescence, the illumination of colorful butterflies and drab snakes, the light of knowledge. I am content.

Now, on my porch, scattered clouds on the horizon yield a

spectacular sunset. The sky is enveloped in shades of liquid gold: greenish-gold, yellow-gold, red-gold, all against the black silhouette of several cecropia trees on the hillside. I sip my wine peacefully and happily, watching abandoned oropendola nests swing gently from some upper branches of the cecropia. The light on my face feels wonderful.

Sunset from the Rara Avis Porch

CHAPTER 7

SUFFICIENT NOT SATISFYING

I don't think we know all about this matter...I think it is better if we do not fill up our ignorance with words.
— *Marston Bates,* The Forest and the Sea

Ways of seeing become ways of knowing. How you choose to see the natural world depends on what you want to know about it, and what you know is conditioned by what you have seen.
— *Janice Emily Bowers,* The Mountains Next Door

Am I hearing a chihuahua barking? I shake myself awake. Remembering where I am, I realize it is a bird. It is loud. I go to the porch. It's still dark, but the sky is noticeably light in the east. The high-pitched bark continues, sharp and brief. It is probably 50 meters inside the jungle. The call ceases once it is fully-light. At breakfast I ask Megan. "A Barred Forest-falcon," she responds. I'm so glad she's here because I probably would not have figured it out.

I heard it. According to the "rules" of the American Birding Association, I could "count" it. I didn't. I need to see it first. Barred Forest-falcons, as their name implies, are ambush hunters of the deep forest. Very hard-to-see. I'm still looking for my first.

My personal rule for my birding life list, is only to "count" birds I've seen. After I've seen a bird well, and can recognize the call on my own, then I will list it on a daily bird list if I hear it only. This means some sightings fall into a category I've coded as SNS: *Sufficient Not Satisfying*. That's how I describe an experience complete enough for me to say I've seen the bird but fell short of what I wanted.

My first sighting of a Black-crowned Antpitta makes a good example. As I described in Chapter 4, the bird is narrowly distributed, therefore rare. It is also very difficult to see even when you are in a locale it inhabits. My first "sighting" was with a birding group I had organized as a fund-raiser for the local Audubon Society. We had one of the best guides in Costa Rica, which says a lot, because there are many high-quality guides. We were on a trail in Braulio Carrillo National Park. Not only was this trail not all that far, as the toucan flies, from Rara Avis, it wound through the same mid-elevation terrain.

Two years before, the same guide on the same trail stopped and announced that he was going to play the call of the Black-crowned Antpitta. The trip mate standing next to me made quick and knowing eye contact. I had spoken of antpittas. She knew how important they were to me. Ernesto played the call a few times. I peered intently into the nearby vegetation. We all listened but knew that Ernesto's hearing was so much better that if there was anything to hear, he would detect it first.

I hoped. I'm sure we all did, but no one more than me. None of us saw or heard anything. Trolling, that is playing a call without knowledge the bird is nearby, rarely works. In reality, when he said we'd "try for it," my expectations were low.

The next year I made the trip to Rara Avis where I encountered the drought and silent birds. A year later, I organized another Audubon group and we were back again in Braulio Carrillo, on the same trail, and a Black-crowned Antpitta had just called. Instead of detecting our group and moving off, it responded to the playing of its territorial call and moved closer. Our duel with this bird lasted 45 minutes. Once, in the middle of it, I broke my concentration and glanced about the group. There were eight of us plus the guide. My wife and one other person, not so interested in the birds, were hiking on their own. The rest of us were frozen in place. It was like a game of "Simon Sez," as we waited for the next command. We knew that antpittas are so shy that any slight movement detected by the bird would send it away, dashing our chances of seeing it.

We were spread out on both sides of a small ravine that opened above us and ended at the trail on which we stood. About half of us were on one side of the ravine, and half on the other. I was in the middle, with a good view of the others merely by using my peripheral vision.

I particularly recall one of my friends as he stared intently at the middle of the ravine. His binoculars were held stiffly just below eye level so he could raise them with a minimum of movement. His face showed such intense concentration, one would assume the matter in front of him was an occasion of life or death. His body position, being on a hill, appeared somewhat awkward, but he was well-braced, determined not to move or shuffle his feet, lest he startle the bird. I thought of an ex-sister-in-law's comment about my brother-in-law while he was intensely fishing. "You'd have thought the dumb things were diamonds," she had said contemptuously, which I now know was probably a hint about their impending break-up. I felt a wave of amusement at my birding friend's intensity until I realized I was the same, if not more so.

This antpitta continued to be fooled by Ernesto's call. It wanted to confront the rival and chase it away. Somehow in the dark understory, Ernesto sensed the bird's movement. "It's there," he hissed. "Where?" I whispered. I couldn't find it. After a few more hisses and an almost imperceptible point with his nose, I finally detected movement under some small plants. The bird blended in so well. It's habitat is so dark that most of the movements gave the impression of a piece of soil moving. Then the bird was gone, only to reappear several yards away. I'd catch the movement in the corner of my eye and bring my binoculars up again--too late.

As I think back, I consider that the bird's uncanny ability to vary its speed and direction is part of its strategy when avoiding predators. Surely, the only way these are eaten is if ambushed. I probably glimpsed part of the bird ten or more times. My best view was of its rear half as it hunched its wings over its back while executing a rapid spin before running up the slope. Only one person managed to have her binoculars at the right spot to glimpse the scaly-breast. I put the bird on my list, but I wasn't satisfied. Our guide told us he had never been able to show one to a group and had only seen it himself twice previously.

My current trip to Rara Avis is already a success. The Lanceolated Monklet and Thicket Antpitta sightings are everything I could hope for, but, deep down, I also know that the Black-crowned Antpitta, because of my history with it, and its rarity, remains the real prize.

So, I was elated when Jason told me the previous evening that an antswarm had passed directly in front of his manakin viewing location. He had seen two Black-crowned Antpittas at close range. What luck!

That's where I am headed early this morning. I am planning to ascend the Margot Trail to reach Guacimo where the ants had been seen, but while still on the tractor road near the Azul trail, I hear a squawk that I recognize as an antbird. I see movement. There on a low branch in front of me is an Ocellated Antbird. Doubtlessly Costa Rica's most handsome antbird, these have a bright blue face highlighted by a black throat and neck. Otherwise the plumage is overall bright rufous but with a black, scaly pattern. I am thrilled. I have only seen these a few times and never so well.

What else might be going on along the Azul Trail?, I think. I feel some reluctance, having an appointment with the ants on Guacimo. Ascending the Margot Trail would be quicker, but I decide to ascend Azul for a few minutes. After about 15 meters, I hear another, different call, a series of brief, squawks that remind me of what one book calls the "shek" of the Steller's Jay of North America's Western Mountains. I know for certain I have just heard a Black-crowned Antpitta, and it is very near.

I play the territorial call, the long series of whistles, and wait. The Black-crowned Antpitta calls back. The battle is joined! I see the bird four times. The first three views are of the chocolate brown shape, a sunlit back view and the tail. At last, I spy the entire bird in my binoculars. The view is from the side. Although I can't see the scaly breast, I can see the head shape, the rich rufous color and the black crown. Just as before at Braulio Carrillo, I never see the bird coming. I am never able to anticipate where it will appear except by sound. When it does appear, it is as if it emerged from a hole in the ground.

What a duel! It calls. I call. We trade squawks too, but mostly duel with the long, territorial call. As with the Thicket Antpitta, the Black-crowned Antpitta doesn't move while it sings. The lengthy song enables me to determine its relative position and move a bit to ensure it doesn't see me.

It moves while I play its song. Sometimes I can hear what I take to be a mad *churr* while it is moving. A few times it comes very close, while managing to stay concealed. A couple of times it encircles me. That's when I feel frustration. The vegetation here is thick, but not so thick that I shouldn't detect it passing me.

Eventually, I recognize its preference for the downhill slope in front of me. I move slightly during one of its songs so that my own calling is more likely to keep it in front of me. I learn its paths, and while it sings, I improve my position in hopes of intercepting it. Those occasions are when I have my views, the best coming after more than an hour. Eventually, I realize it is continuing to encircle me in the same wide circle. It has stopped approaching. It isn't coming closer. *Let's call it a draw,* I think and silently back away with satisfaction at seeing the bird, and the hope that it knows it is still the king of that patch of jungle, having vanquished a sturdy and persistent rival.

I survey the scene as I leave. I have kept low, mostly slinking, and sometimes slithering in the forest litter. *Dumb,* I think. *I'm glad there were no snakes!* There wouldn't have been room for two people to stay concealed, much less a group. No wonder there are so few sightings of Black-crowned Antpittas. I look at my watch. I am amazed. It is 8 AM. An hour and 45 minutes has elapsed. What a gift it was to have been so engrossed in an activity!

Now on to Guacimo and the antswarm. I knew the location and quickly found a remnant of the swarm. A few ants were still crossing the trail. Unfortunately, judging by the sounds, the main cohort of the ants and their attendants were maybe 10 meters away through thick vegetation on a steep hillside. Closer access without tumbling down the slope was impossible. It was disappointing and exasperating to hear the commotion. Maybe there was a Rufous-vented Ground-cuckoo? Maybe the even rarer Wing-banded Antbird, once reported here, was in attendance? I couldn't know! There's so much I could know. The rainforest is an awesome encyclopedia of immense size. Sometimes, as with that antswarm, I have an idea what I'm missing. Usually, I don't have a clue.

Another favorite Costa Rican location for me is the Las Cruces Biological Station. I've walked and looked for birds there many times, often for four or five hours. My bird lists for those walks typically approach 40 species and may include some very special encounters. Examples include the rare Chiriquí Quail-dove which once performed a pirouette in the middle of the trail. Another time, I found a just-fledged Rufous-tailed Jacamar precariously clinging to a muddy bank. I marveled at how well its jewel-like colors blended with the orange-red soil and green leaves.

I describe experiences such as these and my family and many friends believe I'm some sort of genius when it comes to birds. One aspiring birder, not very proficient at the time, described me as "the best" he'd ever seen. I'm sure he would take back that comment now that he's more experienced. Certainly, I know better. Indeed, the night before, I was congratulating myself for my knowledge of birds at Rara Avis. I thought I would share this wisdom with my educated brethren at the dinner table. I remarked "Ochre-bellied Flycatchers are supposed to be common in this habitat, but their absence at Rara Avis shows how so many species in this country have spotty distributions." "That's funny," replied Megan. "That's the most common species we capture in our mist nets." She said this nicely. It wasn't a "put-down." It was simply a matter-of-fact comment. I knew I had been missing many species; now I knew one more.

On a recent trip, I had recorded 32 species in seven hours of birding at Las Cruces. Within the same week, I saw the list from one of the area's professional birders. He had recorded 102 species in less than five hours. I felt disappointed and inadequate. The secret is that the other birder knew every sound in the forest. I knew a few. In comparison to the skills of many others, I don't know anything. So I keep working on it.

Navigating the waters between satisfaction and disappointment is something I find difficult. How does one achieve without aiming high? If one aims too high, yet still does well, is that an experience to be celebrated or treated with dismay because the sought after "mark" was not achieved?

I once had a co-worker who was a dour individual. It was fitting that he had a sign in his office which said, "Blessed are those with low expectations for they shall not be disappointed." I suppose he was never very disappointed, but he also didn't seem very happy. He didn't seem to care very much for his career, for his wife, or for the religion which consumed most of his spare time. As for me, I live with high expectations. If something is wrong, I try to fix it. If I learn that something good could be better, I often try to keep improving. This can result in dissatisfaction even when I've done well.

I once asked my family what sort of annoying habits I had. One was disclaimers I habitually made about my cooking and baking. I like to make artisan bread. I frequently compare my products to some perhaps-unattainable ideal. They were weary of hearing what went wrong with the rise and why the distribution of holes in the bread wasn't ideal, etc. Never satisfied!

I once read how important it is to develop a trust that situations will turn out OK. This way of thinking would lead to greater satisfaction with life. An example the author gave was that "no one would dig up seeds to see how they are doing." Oops! I do exactly that--or, at least, I used to. I like to garden and would worry about how things were doing, and, sure enough, I would often carefully uncover seeds to see if they had sprouted. I solved that problem by seldom planting seeds in my garden anymore. I sprout them externally and then plant them!

Fortunately, the lack of satisfaction does not carry over into everything. I never, ever, doubted my marriage. I can barely put up with me much of the time, and my wife does all the time. No one else would so easily endure my moods and complaints. Best of all, besides her endurance, she never keeps score. I might push her patience to the limit a hundred times to only one of hers. No matter. I know how she does it. Without realizing it, she personifies one of the steps of the Buddha's Eightfold Path: Right Intention. She always means well and trusts that she does. If something doesn't work as well as she had hoped, she doesn't internalize it as a mistake. She doesn't ruminate on what she should have done differently. Her example has made me realize that, to a certain extent, the way to be satisfied is to refuse to be dissatisfied.

Mary is also good at "letting go." There are many awful things that happen personally and on the world stage, and "let it go" has become a New Age mantra. At times of stress and fear in our lives, I've marveled at Mary's ability to enjoy a good meal or laugh at a joke. I couldn't do it. Years went by. I accepted the logic of the need to "let go," as well as the accompanying clarification that "letting go" didn't mean deciding whatever was troubling me was "ok" or "easy-to-endure." I still couldn't do it.

I would see people who had greater struggles than mine who seemed to handle them with a level of equanimity that was impossible in my view. How did they do it? How could I do it?

Even now as I write, I think, *I probably can't do it*. What bad things are out there for me, I can't know. Maybe none. Maybe I'll die in my sleep while my loved ones are healthy and happy, and humans haven't yet destroyed the planet. Maybe I'll be that lucky. Probably, I won't. But by changing the terminology, I've learned what "letting go" really is. For me, the way to say it is, "making room for it." Erroneously, I always tried to "let go" by making the pain or fear or uncertainty go away. I failed and blamed myself. I've learned that those last two sentences are a succinct recipe for depression. "Making room," is something else and ultimately a secret to successful living. There is beauty and satisfaction amidst the pain.

Now I can understand how a dear friend dying of a horrible disease could still write beautiful words. He made room for his death but kept room for other parts of his mind. Recently, I came across these words from the author and holocaust survivor, Primo Levi: "Sooner or later in life everyone discovers that perfect happiness is unrealizable, but there are few who pause to consider the antithesis: that perfect unhappiness is equally unattainable." This is much like a story told of the Dalai Lama. When asked how he could be so cheerful while living with the loss of his country, he remarked that although "they" had taken his country, he wasn't going to give them his mind as well.

Do you remember the old cartoon series, "Calvin and Hobbes?" Calvin being a mostly snotty, yet often insightful six-year old, and Hobbes being his stuffed Tiger, who nonetheless would often philosophize in the manner of his namesake. A favorite of mine was when they fantasized about being able to wish for anything they wanted. Hobbes said he wanted a sandwich. Calvin was incredulous, accusing Hobbes of a "failure of imagination," as he wished for wealth and power. In the last panel, Calvin sat there angry and dissatisfied. Meanwhile, there sat Hobbes happily eating a peanut butter and jelly sandwich. Said Hobbes, "I got *My* wish."

Growing up, I could never choose the sandwich; I had to aim high. Part of it was due to my upbringing. My mother, as I related, never showed much satisfaction with me. "There but for the grace of god!" If I did well, I always heard something to the effect of, "that was good...but..." The hindsight of 50 years allows me to recognize that my mother never had any choices. No wonder she insisted I not squander mine.

It is important to emphasize, Mom meant well but her way of communicating, combined with my way of taking in "her way," was not healthful. Ultimately, I have tried to follow my Dad's example. He lived through two unspeakable tragedies, his experiences in World War II and the horrific cancer-caused decline and death of my Mom at age 50. Despite these tragedies, Dad mostly looked at life as one lucky accident after another. Once, I was trying to convince Mom that she should not save money by buying whole chickens because my brother, she, and I always argued over the white meat. I turned to Dad. "Which part of the chicken do you like the best?" "The white meat," he replied. "But you never ask for it?" I said. "I know," he replied. "Growing up, I only got the neck. Now I get the back and thighs!" Dad would say something like that only when asked and only in a down-to-earth manner with no hint of any intended lesson or admonition.

Over time, I realized how wise my Dad was. "As long as you have your health," he would say. Others of his aphorisms were, "That's part of life," and the family favorite, "You gotta eat!" His father died of cirrhosis of the liver and my mom related that Dad had told her a family Christmas dinner was ruined by his father's drinking the money meant for purchase of a Christmas turkey. Maybe Dad knew more hunger as he grew up than he let on.

At least now, I am, for once, well-satisfied with my birding prowess. I have seen both antpittas I sought, and I have done it on my own.

I can barely contain my smug self-satisfaction at dinner that night. I expect the researchers to congratulate me for the great bird I've seen. Instead, their enthusiasm is muted. Is it because they have all seen the bird already? One saw them in the antswarm. One of the others doesn't seem particularly interested in birds beyond her manakin-watching task at hand. *Well,* I think, *at least I'm satisfied.*

Later, I also wondered if their lack of enthusiasm could have been because I had used a call. Once, I was in a birding group when the leader announced a call was to be played in hopes of enticing a bird to pop up so everyone could get a good look. I watched as one couple put their heads down and slunk away, as if to be sure no one suspected them of participating in a shameful ritual. Meanwhile, interest piqued; others in the group strained to be at the front in hopes the bird would respond.

To me, the controversy of using calls is silly if not disingenuous. Would those who disavow calls stop participating in owl surveys where calling is essential? Would they put away their bird feeders? Would they cease tramping through the rainforest, or the prairie, or the swamp? If you were a bird, what would you do if you noticed someone pointing two big, circular, black eyes at you? You would bolt, as do many birds. Be smart. Make sure you aren't disrupting the bird's behavior unnecessarily. And, recognize that for some heavily birded areas and much sought-after birds where calling would be incessant, don't use a call.

Feeding birds in your yard might seem innocuous, but close concentrations of too many birds can lead to disease. Another consequence of so much bird feeding in the United States is an apparent increase in sightings and probably population of the Cooper's Hawk. What do Cooper's Hawks eat? Other birds. Rare is the bird feeder that isn't frequented, at least occasionally, by one of these fast-flying predators. Birds, like humans, are attracted by an easy living. What could look better to a Cooper's Hawk than bird feeders that concentrate its prey?

Nonetheless, I enjoy the bird feeders in my yard, but just as my friend Alejandro is not enamored with feeding antpittas, I am far more satisfied with my success at seeing the antpittas on my own rather than at a feeding station. The most fun and satisfaction from viewing birds comes from understanding their life histories and being able to watch their normal behaviors.

Here's a personal example. The Le Conte's Thrasher is a rare bird in the United States. Their populations are in real trouble. The group Partners-In-Flight has them on their "red watch list" because of their restricted range and steep population decline in recent decades. Probably never very common, their bad luck is to live in flat desert areas that, if not desirable for housing, make perfect playgrounds for off-road vehicles and target shooting. They live in such a sparse habitat that population densities are naturally low.

When I lived in Tucson, Arizona, Le Conte's Thrashers were reported west of the city in the sparsest deserts, but they were seen so rarely that I considered them near mythical. The internet, however, has enabled so much birder communication that it wasn't long until an area near Phoenix became known as a good spot to find the bird. There were many reports of birders showing up in the springtime, playing a call and having the bird pop up on a bush and look around for the intruder. Sometimes, if the birder was lucky during breeding season, there would be thrashers singing from exposed perches early in the morning, offering long views with spotting scopes. I arranged to visit.

I spent five hours beginning just after sunrise and neither saw nor heard a Le Conte's thrasher. I walked all over the supposed "best" location. I played the call several times to see if one would come out of hiding to challenge an intruder. Nothing. It wasn't as if I was alone either. I had joined up with a birder from Wisconsin who had made a special trip just for this bird. Distantly, we could see a couple of others, all hoping to see the Le Conte's Thrasher. Together, my companion and I crisscrossed the area finding a few other species, but no Le Conte's. My time ran out, and I had to leave. Later, I learned that my partner had successfully seen the bird several hours later. Under what circumstances, that is, whether he continued to use a call, I don't know.

Still bereft of a Le Conte's Thrasher sighting, I returned the following January. Once again, I had aligned my trip with the birds' mating season, hoping they would be active and easier to see. I had planned another dawn visit, but some prior social obligations ended earlier than expected, and I had a few extra hours for the late afternoon. I decided to reconnoiter the site to improve my morning quest for the bird. I didn't want to just "troll" for it by walking around and randomly playing the call. I had done a bit more research regarding its habits, and I wanted to see if I could recognize the best places to find one. I honestly had no expectation of seeing one that late afternoon. The bird is known for its secretive habits. I was visiting near the end of the day. It wasn't going to do any good to play the call so near dusk.

As I walked, I came to a likely looking area. While making a mental note to find the location in the morning, I saw some movement. Slowly, I raised my binoculars, and there was a foot-long, sandy-colored bird walking among the low bushes—a Le Conte's Thrasher. But it was acting strangely. It was raising its wings up and down and fanning its tail. This was nice because it was clearly showing the tawny-colored vent area, one of the principal field marks. Recognition dawned! It was displaying! It's wings were partially extended as if covering chicks underneath. The tail was raised to show the vent area, the only color on the bird that doesn't match the brown of desert soil. Using its legs, it performed squats followed by a couple of pirouettes. I scanned the nearby bushes and found a female, watching intently.

The female hopped to another bush. The male followed continuing his dance. I watched this spectacle for ten or more minutes before the birds moved from view to where I would probably spook them if I followed. Yes, that is a long-story about a bird that is mostly the color of Arizona desert soil, but the point is that if I had been willy-nilly playing a call in hopes of causing one to hop up on a bush, I would never have seen the courtship behavior. Of all my bird sightings, that one makes my top ten list, because of the rarity of the bird as well as the behavior I witnessed. How satisfying!

Yet again, it is all relative isn't it? Some folks save for their entire lives and make a single trip to a location in Costa Rica, where a collection of hummingbird feeders might yield more than ten species in a matter of minutes. Their satisfaction at having the money, the time, and then the experience of watching exotic species may exceed mine from days of searching for antpittas.

Am I jaded? I've been to Costa Rica nearly thirty times. I keep wanting more. *Maybe that's ok,* I tell myself. *What's important is not to need more.* Jon Kabat-Zin, one of the leaders in popularizing secular Buddhism in the West, cautions about being greedy—always wanting more. It is a trap I often find myself in. As Kabat-Zin notes, we can find ourselves alternately shifting between "pursuit of what we like and flight from what we don't like. Such a course will lead to few moments of peace or happiness. How could it? There will always be cause for anxiety. At any moment you might lose what you already have. Or you might never get what you want. Or you might get it and find out it wasn't what you wanted after all."

Back to Calvin and Hobbes: Calvin once defeated Hobbes in a game of checkers. In the second panel he says, "I'm the champion! I'm the best there is! I'm the top of the heap!" He surveys his victory with a self-satisfied look in panel three. In panel four he looks miserable. "Is this all there is?"

These days, with the experience of many years, I don't make Calvin's mistake as much. I stop and revel in the present moment. I realize how fleeting is any success or good feeling. For a long time, I resisted that sense of impermanence. With much practice, however, I have experienced that if I can be fully "present" for the success, I will enjoy it more and more easily accept how soon it vanishes.

I also realize that instead of "wanting" more, or always trying to "make it better," I need to make a semantic change. It is ok to "try" for more as long as I don't stake my happiness on the outcome. In that way, it is the pursuit that's fun. Here again, I can see the A+ on my written exam. I have impeccably answered the essay question regarding the problems of desire and aversion and the Buddhist response. How does it go in my day-to-day life? Sufficient (often), not (always) satisfying!

CHAPTER **8**

YOU DON'T BELONG HERE

A man said to the universe:
'Sir, I exist!'
'However,' replied the universe,
'The fact has not created in me
A sense of obligation.'
— *Stephen Crane,* "A Man Said to the Universe"

So often we stride through the world, thoughtlessly claiming it as our own. It is good for us to feel that we have no right. This is another blessing of wilderness—to remind us that we intrude.
— *Janice Emily Bowers,* The Mountains Next Door

Ah life. The trip has been beyond successful. That's what I was thinking when I went to bed last night.

The next thing I remember is a searing pain in a finger. Perhaps you've dreamed of pain and even woken from it.

But this is no dream. I leap from the bed and there on the floor, teeth bared and hissing, is a large rat. The next thing I know, I am standing on my bed, heart racing. The rat runs under the bed. I jump off and grab my tripod to chase it or kill it or do whatever I need to do. I look under the bed. The rat gnashes its teeth and hisses again before darting around the corner and into the night. I look at my finger. Blood. Not good. If it hadn't broken the skin, I could have easily forgotten it.

I know what has bitten me—a Watson's Climbing Rat. They are well-known at Rara Avis. One of the field guides to Costa Rican mammals even mentions the Rara Avis hotel as one of the best places to see one. It is 3:30, still nearly two hours until breakfast. I inspect my finger. Shallow punctures on both sides ooze some blood. I wash it with soap and douse it with antiseptic and then repeat for good measure. What sleep I manage afterwards is fitful at best.

The incident creates consternation at breakfast. The biologists all say they have been given rabies immunization before the field work. They also tell me they have been sleeping under bug nets. None of that helps me. After a few minutes of consultation, Megan says she has the ability to give the shots. Maybe the rabies dose can be sent up with the tractor and I can stay. I don't want to leave because there would not be time to go out, get a shot, and return. I would simply end up with three lost days. Giancarlo finally suggests that I call Amos Bien, the proprietor.

My plan for the day, of course, is history. What I feel is intense aggravation. If the rat is rabid, I will be able easily to start shots on time, so I am not worried about getting sick, only about my trip being truncated.

I have to wait for a couple of hours before we can reach Amos. The call goes like this:

> Me: "Uh, hi Amos, I was bitten last night on the finger by a Watson's Climbing Rat."
>
> Amos: "They are always all over the hotel. In more than thirty years, no one has ever been bitten. Did you corner it or how did you annoy it?"
>
> Me: "I didn't bother it. It bit me while I was sound asleep."
>
> Amos: *Long pause*. "Hmm. That's bad! I'll call a doctor and we'll figure out what to do."

Suddenly, my memory is jogged by a long-ago event. "When you are talking to the doctor," I say, "be sure and tell him that about 35 years ago, I was bitten by a dog during a rabies epidemic in Arizona. The owner of the dog gave me a false name and address. I had to undergo the complete rabies shot regimen at that time." Back then, the scenario was fifteen shots in the stomach, given circularly, once per day around the navel. The memory of how much some of those shots hurt persists in my mind. Scar tissue developed from a couple of them, leaving hard nodules in my belly that persisted for years. I never thought the incident during the rabies epidemic would have any sort of positive outcome, but it does. When Amos calls back, he says the doctor told him what I had undergone was the equivalent of the basic immunization given veterinarians and field biologists. All I need is a couple of boosters. Viviana, his office assistant, will pick me up the day I return to the city and take me to a hospital for the first one. I just have to arrange for one other back in the US.

Tragedy is averted, but my day is compromised. If I had it to do over again, I would return to my room and meditate for thirty minutes before heading to the jungle. Instead, tired and emotionally drained, I go into the jungle and try to make up for lost time.

It doesn't work.

When things go wrong, I struggle to regain balance and, instead, focus on any other imperfection I can find. I had been feeling harmony with the jungle, and now I dwell only on the incidents of alienation. I could have considered Skutch, who often wrote of the harmony of the rainforest. He loved the sight of multiple species feeding at the same fruiting tray. There would be various multi-colored tanagers, honeycreepers, and saltators. Sometimes a woodpecker would drop in, and even though they temporarily chased the others, the Fiery-billed Aracaris (Los Cusingos), the small toucans for which he named his farm, didn't seem to strike much fear in the other birds.

He saw the rainforest as a near Garden of Eden, where trees would provide fruit for beautiful birds that would harmoniously share the tree with their brethren and fly off propagating the next generation of trees through the seeds in their droppings. On the other hand, Skutch's desire for harmony caused him to revile most predators, especially the deadliest despoiler of bird's nests, snakes. These he would kill when they appeared near his home, but otherwise he maintained a peaceful co-existence with almost everything else.

I have observed that in his voluminous writings about tropical birds, there is very little about birds of prey. In my own life, I have done a great deal of work with owls. Their lack of mention in Skutch's work is conspicuous but telling. In *Merenda,* his only full-length book of fiction, there is a passage where the namesake character, who lives in remarkable harmony with nature, confesses her fear of the dark. When asked what she thinks about at night, she recounts visions of snakes who creep "through the bushes trying to surprise birds slumbering...weasels prowl stealthily...ready to pounce...vampire bats flit silently ...seeking living victims of their gory feast." Finally, she comes to owls and relates how they are "so different from other birds...flying through the night on noiseless wings, hunting mice and small birds."

Presumably, as with most predators, Skutch didn't care for owls. There was one exception to his revulsion of predation, the Laughing Falcon, called *guaco* by the locals for its call. *Guacos* earned Skutch's admiration because of their principal prey item: snakes.

As I noted in chapter 3, I admire Skutch for the equanimity that fills his writings. A friend of his described him in "good spirits" even though nearly deaf and blind and a few weeks before his death, which occurred a week before his 100th-birthday. In any case, this day, I don't think of Skutch. I begin to ruminate about what else has gone awry.

On the day of my hike on the Levi trail, I had stopped by a broad stream. This is the only stream of consequence in this area of the reserve, in contrast to the steep and wet area I described in Chapter 4. Not only is the stream broad, but there are a series of rock slabs which have created a large light gap, perfect for sunning oneself. At the time, I was proud of myself. I hadn't let the warm, sunny day and consequent slow birding bother me. I had gloried in the butterflies, the snakes, the lizards, and interesting plants. Now I would take the opportunity to bathe in this beautiful stream and dry out on the rocks. Life was perfect!

I removed one boot and sock and while I was removing the other, a bee landed on my bare foot. It stung me! It immediately flew away. My foot hurt. I plunged both feet into the creek to soak them, both for hiking relief and to dull the sting. The pain subsided after a few minutes, leaving me with only an itch. I felt betrayed by the sting. *I'm here loving this place, and this is the welcome I get. How unfair!* Not knowing if it had partners nearby, I abandoned my bathing idea, and sat there fuming.

The sting was a trivial incident, but a reminder that nothing we can do protects us from the realities of life. That's why I selected the Stephen Crane poem "A Man Said to the Universe" for the introduction to this chapter. I can't say Crane's poem is a favorite. In the same way, I wonder if any Buddhist can be said to "love" the First Noble Truth often presented as "Life is suffering." Love it or not, its accuracy is undeniable. As Paul Simon, whom I consider my generation's finest poet, once sang, "Everything put together sooner or later falls apart."

When things go wrong, instead of embracing equanimity by recalling the First Noble Truth, my mind typically becomes very busy trying to understand and explain what happened.

I was angry about my rat bite. You might be thinking, "At least it wasn't a snake!" While true, I have learned from my own injuries and convalescence and subsequent reading on the subject that when someone has suffered a setback, and you wish to comfort them, you are making a mistake if you begin any sentence with "at least." People who are hurting need empathy, a sense that you, the comforter, know how they feel. Unfortunately, I wasn't giving myself empathy either.

The rat bite was bad luck, but I also know it was my fault. Rara Avis supplies each room with a large glass jar for any foodstuffs one might bring, while at the same time suggesting that you not bring any. That wouldn't work for me. I needed ten days of lunches because of the long hikes I was taking. I also had a few snacks along, knowing there were no "stores" in the reserve. My backpack smelled of food, and so did my clothing. I had food in my pockets and crumbs on myself, and probably wiped some peanut butter or frijoles on my pants during lunch. The hotel was otherwise empty; I was the only attraction in town for the hotel rats. I should have stored my food and soiled gear in another room. *How stupid,* I tell myself. Now I am in familiar territory. Not only has something bad happened, but I've internalized it. It is all my fault! Now I can add beating myself up to the problems that have already accumulated. Buddhists use a metaphor for this when they speak of the warrior who is hit by an arrow. Instead of simply pulling it out, he keeps twisting it. That is me.

I also blame myself for wearing filthy clothes. It might seem counter-intuitive, but it makes sense to wash clothes less often in this type of environment. The humidity is so high and mud so prevalent that the state of one's clothing after one hour in the field or twenty hours isn't much different. Another issue is that washed clothes never seem to dry. Talking to the researchers, they were doing the same as I, that is, showering after work and changing into clean clothes in the evening, and then putting on the same work clothes day after day. Kim even told me she'd worn the same pair of pants for more than a week.

I was changing every two days, but there was an important difference. Not only was I staying in the field longer, but, on most days, I was doing some night hiking. I didn't want to take multiple cold showers or be changing my clothes back and forth. I wore the dirty clothes until after dinner when I would do a short night-walk and look for frogs and snakes and call for owls. I was in my clothes from 4:45 AM until after 8 PM. The others were probably out of their work clothes at 4 PM. They spent most of their day sitting on a rock or stool. Their work clothes didn't get as dirty and had more time to dry out, which probably limited the amount of vermin they had collected.

My dirty clothing led to additional troubles which surfaced after the trip. The day before the morning of the rat bite was the day I saw the Black-crowned Antpitta. Seeing the Thicket Antpitta, my first antpitta success, was still the highlight of the trip, but the Black-crowned Antpitta had been more of a challenge. The Black-crowned was more reticent to venture close. It mostly moved back and forth in an arc in a small ravine in front of me. I didn't find anywhere I could conceal myself while maintaining a viewing area. So, I slithered back and forth behind a downed log. Described another way, I rolled intimately in rotting leaf litter in a tropical rainforest for more than an hour. It isn't a stretch to say that by the time I finally stood up, I probably had creatures on me that had never been identified.

After seeing the Black-crowned Antpitta, I went on with my day. That might have been ok, but I had already been planning a night hike. In hindsight, I remember the clothing feeling itchy. I also found a few creatures crawling about, some very tiny red mites or ticks that I brushed off. On that day and the next, however, I don't recall receiving bites beyond the ever-present mosquitoes.

Insect bites are basically a given in the jungle. I had pre-treated my clothing with Permethrin, and I was wearing tall boots, both standard practices for minimizing bites. I had DEET and was using it when I had insects flying about my exposed flesh (face and hands). Nevertheless, I always received bites, and the partially effective standard practices were not designed for rolling in damp soil and leaf litter. Upon my return to the US, a few of the bites, probably 15 to 20, although small, seemed to be especially bothersome. I had them on both wrists, on my ankles and lower legs, and a couple on my lower left side, just below the ribs—all locations either uncovered, or where there were convenient openings for entry. Previous experience has taught me never to scratch bites. I've gotten rather good at "mind over matter," and "letting them go." It wasn't long, however, before I had purchased some anti-itch over-the-counter salves. These tended to work well—for about three minutes. About the time I was considering going to a doctor, the many itches subsided.

A few weeks later, however, many of them re-appeared, and were as bothersome as in the beginning. Instead of 15 to 20, there were now 10 to 12. This cycle repeated at about 6 to 8 week intervals for more than a year, until only those on my left side remained. I did internet searches and found nothing definitive. There were, however, others who described exactly what I was going through, who had zero success with their doctors. One of them had unsuccessfully begged the doctor to simply excise the offending tissue. I should say that these recurring bites were not like those of a mosquito, but had a much sharper pain. It felt as if I was being stuck by needles. Finally, more than a year after my trip, I went to my family doctor. Biopsy results were inconclusive. I eventually went to a former military doctor and dermatologist who specialized in tropical medicine. She had no idea. After a couple of years, the recurrences began to be less severe, and have finally, I hope, stopped.

That long-term itch, whenever it returned, reminded me of the trip, seeing the antpittas and how much it all meant to me. But it also reminded me that I didn't belong there. I could experience the rainforest, but it wasn't something a person with my life skills could belong to.

Anyway, on this day, I keep beating myself up. *This would have been another perfect day if I'd taken proper care of my belongings,* I tell myself. From the interactions with the researchers and the phone calls, I was in my head most of that morning. My sense of belonging and communing with the forest felt like a distant memory.

Heavy rain in the afternoon doesn't help. I have my raincoat but know that the sauna effect of wearing it would compound my misery. Frantically, I search for shelter and finally find some thick vegetation hanging over a slightly leaning trunk. If I don't move, I am out of the rain. It isn't comfortable, nor is it completely dry. I want to be birding. It is wet and dark. *What am I doing here?* I think. *I could be home, warm and safe. Soon it will be dinner time and I could have a glass of wine and nice conversation with my wife. Why did I ever think this trip was a good idea?*

The rain stops after a while, but the sky never lightens. The jungle shade enhances the dark sky. Birds are active, as they often are after a rain, but all I can perceive are shapes in the tops of trees. It makes for a less-than-satisfying afternoon.

My best sighting of the day is of a Golden-bellied Flycatcher. It has caught a large cicada, but with its shorter beak, it cannot beat the cicada on the branches to remove the wings, as I've seen jacamars do. The flycatcher tosses the dead cicada in the air over and over but is not satisfied with the catch. Finally, the cicada is positioned just right, and the flycatcher downs it, wings and all.

Watching the destruction of the cicada is distracting and fun, but then Giancarlo calls me on my radio. The radios were a safety feature I had brought along, knowing I would be hiking alone. Ostensibly, he is calling to see if they worked, but more likely, I am on his mind because of the morning's problems. I misunderstand and think he wants me to come in. That makes me anxious, and the timing is awful. A mixed flock has just moved into a fruiting tree in front of me. It is full of birds.

I reluctantly leave the birds and race back. After clearing up that Giancarlo doesn't need to see me, I return quickly to where I have seen the mixed flock, but the birds are gone. It begins again to rain heavily. I do see three curassows and once again think I hear White-crowned Manakins. Right about then, I am careless when pulling my boot from some mud, lose my balance, and do a face plant. No harm done, but not very good for the rhythm of my day.

I return to my room and find fresh rat feces in one corner. My consternation builds some more. I clean up the mess and sit on the porch. I take deep breaths and simply try to be present.

It is a few minutes before five. It has rained very hard and is now so foggy and misty that I cannot see the open-air kitchen where we take our meals, even though it is not more than 30 meters away. The only sound is the *tink, tink, tink* of a tiny rain frog, commonly called—what else?—a tink frog. A slight breeze rustles the fog and mist. This combination of sight, sound and moisture, which I've experienced so many times, soothes me.

A shower adds to my relaxation. Thinking back through the day, I feel good. We have a plan for my rabies inoculations and my trip will not be shortened. I enjoy supper communing with the group, today enhanced by Megan's Phd advisor, Emily, her husband Eliot and their 8-month old son, Will. The conversation is lively and distracting. Peacefully, I return to my room and find rat feces on my newly changed bed.

I am angry more than anxious! *Why me? I love this place!* I remind myself of one of the biggest lessons I know of life: *You don't get what you deserve, you get what happens to you.*

I rummage around the other rooms looking for clean bedding. I find some folded in another room and decide to make the bed and sleep there. It is a restless night. There is no rat, but twice in the night I feel a hot sting. Each time, I find one of those small red ticks or mites. I pluck it off and wash and use antiseptic. The message has been delivered: I can experience this place, I can even wallow in its beauty and mystery, but I do not belong here.

CHAPTER **9**

RECONCILIATION

> *I leave Sisyphus at the foot of the mountain! One always finds one's burden again. But Sisyphus teaches the higher fidelity that negates the gods and raises rocks. He too concludes that all is well. This universe henceforth without a master seems to him neither sterile nor futile.... The struggle itself toward the heights is enough to fill a man's heart. One must imagine Sisyphus happy.*
>
> — *Albert Camus,* "The Myth of Sisyphus"

Yesterday, the jungle drew blood, my blood. Paradise and tranquility were replaced by frustration and anxiety. *That's what life is,* I reminded myself.

And so, it happens this morning, I'm greeted with delight! On their way to breakfast, Megan and Jason caught a caecilian—an amphibian resembling a big worm—ten inches long and purplish. Megan deemed it the slimiest thing she has ever touched. According to Wikipedia, "Caecilians are...limbless, serpentine.... They mostly live hidden in the ground and in stream substrates, making them the least familiar order of amphibians." Accurate that! I pride myself on more-than-usual knowledge of natural history and I'd never heard of them. I examine the caecilian's purple/lavender color as it slithers about in a plastic bag. Thanks to the jungle (and expert biologists) for helping reconcile my frame of mind!

This was a Purple Caecilian (*gymnopsis multiplicada*). As I studied them later, I wondered if author Frank Herbert hadn't used caecilians as the design for the gargantuan worms in his famous sci-fi book, *Dune*. Those worms could propel themselves underground, as do caecilians. Locomotion is accomplished by anchoring the hind parts, forcing the head forward, and then pulling the rest of the body up to reach the head, as in waves. More remarkable, considering their underground lifestyle, is that caecilians copulate and bear live young instead of eggs. (In most amphibians, fertilization occurs outside of the female's body.)

It is barely daylight and I am exhilarated with new knowledge and appreciation of a rare jungle creature. Yesterday's frustration and feelings of estrangement are already memories. My heart rate slows. I look around and feel at home again. I can't wait to begin my hike.

After breakfast, while hiking, I realize yesterday's mood isn't altogether gone. As I briefly revel in the re-found feeling of belonging, my mind wanders to all there is for which I seek reconciliation. I am convinced that some of the species I am seeing will be extinct before my grandchildren reach my age. Their quality of life won't match mine. How do I reconcile myself to such discouraging thoughts?

The brilliant evolutionary theorist Lynn Margulis didn't worry about loss of biodiversity and an eventual collapse of human society. In her view, humans have no more chance of avoiding overpopulation and self-destruction than do fruit flies in a Petri dish (Page 35, *The Wizard and the Prophet* by Charles Mann). Consistent with being co-creator of the Gaia Hypothesis, she believed that no matter what we do, conditions for planetary life (not ours, but somebody's or something's) will be maintained. Maybe I shouldn't worry about it either.

As a naïve young scientist, I was involved in some very early work regarding CFCs (chlorofluorocarbons) in the atmosphere. These are the chemicals that created the ozone hole and were eventually regulated by the Montreal Protocol which became effective in 1989. This example of international cooperation has often been hailed as a model for future planetary issues such as climate change. Except...I worked on the problem in 1971. The science was already far along by then. Eighteen more years were required for a treaty to be established. Moreover, here in 2019, there are news reports of continued violations of the protocol in several countries.

CFCs are a small group of chemicals affecting mostly a single industry. Regulating them was trivial compared to the complexity of climate change. I sympathize with author Jonathan Franzen who has suggested that we would be better off to give up on what will be a futile effort to ameliorate climate change and, instead, focus all efforts on saving as much habitat as possible. His viewpoint has endured much criticism but considering that effects of human/industrial activities on weather were recognized by Alexander von Humboldt as far back as 1800 (*The Invention of Nature: Alexander von Humboldt's*

New World by Andrea Wulf), there's little evidence to suggest humans will take sufficient steps to mitigate the problems.

I'm old enough to recall when many scientists thought that climate change induced by atmospheric emissions could be prevented. Sadly, oilman George W. Bush was elected in 2000 when moderate efforts might have avoided catastrophe. I have always felt that Bill Clinton's dalliance with an intern was key to that election. That brief and stupid sexual encounter was the first nail in our climate's coffin. Twenty years later, President Donald Trump has stated that climate change is a hoax and that the US will not participate in accords, signed in Paris, that recognize it is already too late to prevent two degrees of average global warming. Twenty years from now, I fear that it will be four or five degrees that we will be arguing about. That's why Franzen's point about preserving habitat resonates with me. We are on track to lose both the climate battle and biodiversity. At least if we preserve habitat, adaptable species have reserves in which to survive and to acclimate.

The ecologist and philosopher Garrett Hardin's famous essay on the "Tragedy of the Commons" explains why humans will not address the climate issue. Hardin showed, with a simple mathematical proof, that giving humans a "commons," in climate's case the atmosphere, to use for their shared personal benefit, inevitably leads to ruin. The oft-used argument that "voluntary restrictions" are adequate because it is in everyone's self-interest simply doesn't work. For example, one method to slow the depletion of the Ogallala Aquifer in the US was to require center-pivot irrigation equipment that didn't waste as much water. Squabbling among the users resulted in "voluntary restrictions." What was the result? Some water-users kept irrigating the old way so they could make more money than their neighbors (*Ogallala: Water for a Dry Land* by John Opie).

Later, Hardin modified his essay to say that a "regulated" commons could endure, but these days, uttering the word "regulation" too often brings scorn. Fixing the climate problem, as summarized by Naomi Klein in *This Changes Everything,* not only requires massive international cooperation; it also requires essentially a new economic system. Not going to happen!

Once again, I believe Skutch was correct when he said cooperation, along with appreciation, were the only ideals over which humans could successfully unite. And yet, despite the hopefulness of his words, he was adept at reconciling himself with the inevitable change and destruction occurring all about him. His most personal goal was to learn as much as he could about other creatures without harming them. He accepted the fact that some failures were inevitable but stated that achieving even a few minor successes would be "intensely satisfying." Everything I know about Skutch indicates he never abandoned his stern principles. Although not a Buddhist, Skutch's ability to be reconciled with reality demonstrates an attainment of equanimity for which many have striven but few have achieved.

In contrast, I don't find equanimity in the writings of Margulis and Hardin. I find a clear-eyed view of human nature—a reality demonstrated by a brief article in the June 2018 issue of *National Geographic.* In the article, a professor relates how he tells his students he will give them two points of extra credit if they request it. The wrinkle is that students can obtain up to six points of extra credit as long as no more than 10% of the class asks for six rather than two. Over many years and dozens of classes, only once has he had to give any extra credit. Human greed is ultimately the problem, isn't it? It is my problem too. I was crushed by the disruption caused by yesterday's rat bite, trivial though it was in the overall scheme of things. I felt singled out and wronged.

I've always wished I could be more like my Dad, who, like Skutch, was able to maintain a well-adapted perspective. "As long as you've got your health!" He had an innate understanding of what the philosopher John Dewey called "all the rhythmic crises that punctuate the stream of living." Indeed, as Maria Popova adds, "Our creaturely destiny is intimately entwined with the realities of nature, and nature is forever oscillating between mutually necessary highs and lows" (Brain Pickings: Art As Experience, John Dewey, 2/11/16).

My mind muddles through these thoughts until I become aware of how my sense of being wronged colored so much of yesterday. *Enough of this,* I think. *Focus on today! Focus on this hike*. I've planned a lengthy one—all the way to Plástico and back.

As usual, the thick, overgrown trails make birding difficult. Over and over, I am within a mixed flock, or I encounter a group of noisy flock-leading Carmiol's Tanagers or I hear a calling bird so close it seems I should be able to touch it—and I see none of them. Soon, however, some marvelous sights restore harmony to my psyche.

About to cross a creek, I stop and look about. I continue to hope, in vain as it turns out, that I might encounter a Dull-mantled Antbird. Every time I come to a creek; I play their call. There is never a response, but today at one small creek, after I stand silently listening for several minutes, a Tamandua or Lesser Anteater, comes down the trail. It nearly steps on me!

Tamanduas are handsome little animals. The length, if not the height, of a medium-sized dog, their overall color reminds me of Costa Rican *café con leche*, which contains more milk than coffee. Down the Tamandua's back are two creamy white stripes. Close encounters in the jungle are relatively common if one is still. They have very poor eyesight and rely on their hearing and sense of smell to find prey, mostly ants and termites, and to detect enemies.

The Tamandua sniffs, recognizes something alien, and ambles off. I take another step and flush a rufous brown bird with pale blue skin about its eye—a female Zeledon's Antbird. My usual thrill at seeing a species of the dark understory flows through me. What a great day and a great place!

The trail shifts from the creek's little valley to an area of thick, upland jungle. Movement up in the canopy! Very smooth movement. Something five-armed! A Geoffroy's Spider Monkey is using its four limbs and prehensile tail to swing through the branches. As soon as it spies me, it stops and chatters angrily. Through my binoculars, it's expression seems of anguished rage. One arm raised above the other, the monkey faces me and gnashes its teeth as it sputters. Then, something amazing! The monkey swings itself up to the next tree and the branch it reaches for snaps. The monkey tumbles at least ten meters before grabbing another branch to break its fall. It scampers to the next tree and out of sight.

That was painful!, I think. A human falling that far would have been badly hurt. Of course, the monkey is aided by its smaller body-mass, but still I suspect it went off somewhere and licked some cuts and bruises.

This is exciting. What next? Something on the trail! I suspect a Slaty-breasted Tinamou, which seemed common on previous visits, but here is its larger cousin, the Great Tinamou. Tinamous are one of the most ancient living groups of birds, first appearing in the fossil record in the Miocene (23 to 5 million years ago). I can't think of an analogue for tinamous in North America. Almost flightless, and as large as a North American Sage Grouse, this tinamou is a jungle specialist. Being a specialist, the tinamou is small-brained, thriving only in pristine habitats.

Although depicted as brown in guidebooks, Great Tinamous appear elephant-gray when observed, as they usually are, moving about in the dark jungle understory. They are oddly shaped, like a mostly deflated basketball. The head is turkey-like, but with a shorter neck and longer bill. Big feet and stout gray legs propel it when spotted. Their singular feature, however, is their quavering high-pitched call. Given most often at dawn and dusk, and oftentimes throughout the night, it is a quintessential sound of a pristine rainforest.

The tinamou doesn't seem surprised to see me and slowly walks away. *Strange*, I think. These birds, whenever I had found them 20 to 30 years ago, always blasted away as if about to be shot. That's the difference, I realize. Just as with the many Great Curassows I have been seeing, Great Tinamous at Rara Avis have lived several generations without being persecuted. They've become tame. Almost on cue, I round a corner and there struts a male Great Curassow, decked out in rich, velvet black with its curly crest and bright yellow-based bill. I have seen curassows before at Rara Avis, but never so often as on this trip.

The next interesting sighting is at the beautiful creek that crosses the Atajo trail. Except for those on Platanilla (Chapter 4), this is the only substantial creek crossed by the trails on which there is a relatively long stretch with low gradient. This is where I was stung by a bee last week. Today a Green Kingfisher is present—big-headed, like a sparrow with a heron's bill, according to one guidebook. Crested, spotted, and bright green, they eat fish and aquatic insects and are found from South Texas to Argentina. No doubt their preferred habitat of pristine, clear streams is declining, but hopefully, their large range offers enough protection.

What a beautiful spot. I am tempted to eat lunch, but I have a destination, Plástico. Plástico has always been a landmark on the way to Rara Avis, because often, as described in Chapter 2, the remaining road is impassable or, at minimum, very rough and loud. The first time we were here, my children were 9 and 11. This was well before my daughter became bilingual and the rest of us obtained passing knowledge of Spanish. The tractor driver and his helper spoke no English. All we knew is they had motioned for us to get off the trailer and walk. None of the workers came with us. We didn't know the terrain, nor how far. Although we completed the hike just fine, it was much farther than we expected, taking almost two hours. About the time we were wondering if we were lost or if there really was a lodge, we came upon the spectacular double waterfall and then the tractor road once again, then the bridge and the lodge buildings.

As I arrive at Plástico today, I am thinking that some sunshine would be welcome. I am hoping to see some open-country birds such as flycatchers. Instead, it rains. Although there is usually a caretaker, I find no one about. I can't help but notice how the grounds and everything at Plástico seem well-maintained...solar water heaters...etc. Of course, a tractor could reliably reach Plástico nearly always, unlike the remaining few miles to Rara Avis, which are sometimes impassable.

Eventually the caretaker appears, and we speak briefly. Then I take shelter on the porch of one of the cabins and have lunch. The rain pauses and a few of the "sun" birds are out—common ones such as Great Kiskadee and Golden-hooded Tanager that I add to my day list. An Eastern Kingbird perches at the top of a tree. My mind flashes to playing golf with my Dad on a July day in Illinois. Here it is mid-April; maybe this bird will be on that golf course in a few weeks. It is a small and inter-connected world. Globalization has been part of avian life for millennia.

I enjoy identifying about ten species in five minutes—much different than the difficult forest birding. I had hoped to see raptors but none are flying over the distant hills. There is little activity overall. I return to the tractor road and, again, to the Atajo trail.

The rain ceases and I find myself once again at the creek. A large bird is slowly flapping over the water. I'm thrilled, thinking I am about to identify a Sunbittern—the beautiful, heron-like bird of pristine waterways. I haven't seen one in Costa Rica. It isn't a Sunbittern, however, but rather a Fasciated Tiger-heron. As it sails past, I admire its gray body, bordered by a cream-colored rim to the wings. It lands downstream just in view, gives me one good look, and flies away.

I continue on the trail and encounter a strawberry poison dart frog. This one appears to have blue shorts. I feel rewarded: a tamandua, an antbird, a tinamou, a curassow, a spider monkey, a tiger-heron, and a poison dart frog—all in one day's walk.

Female Great Curassow

There is more. A small brown snake, a Barred Forest-racer, slithers off the path in front of me. I hear, then see a Scale-crested Pygmy-tyrant. The long, complicated name belies the fact that the bird is tiny. This two-to-three-inch flycatcher makes a metallic "clink" sound that can be ubiquitous. It also has a small scaly crest which it erects when displaying. The *peeta-peeta-peeta* of Green Shrike-vireos accompany me as I walk along, and I am frequently reminded of the presence of Carmiol's Tanagers. Their squawks seem omnipresent, designed to feed my frustration. They are the most obvious bird in the forest, yet so difficult to see. I don't recall even hearing them on my initial trips to Rara Avis.

I arrive back at the hotel with substantial daylight. I hear screaming from the nearby rustic casitas. Eight-month old Will is lucky he won't remember his trip to Rara Avis because he spent much of it feverish and teething. I recall time with my own children when they were sick and teething. I imagine how it would be in the sweltering heat and humidity of the very rustic dwelling in which the family is staying. The casita consists of one small room without coverings on the windows. Not only was the humidity near 100%, there was no refrigeration so no cool teething rings. There was no electricity, so nothing for personal distraction or even soft music while young Will screamed and whimpered. Today happened to be Eliot's "turn" while Emily joined Megan in the field. When I asked Elliot about it later, he matter-of-factly declared it was "the worst day of my life."

I decide there is still time to walk in the forest. I retrieve my scope to check out a tree that had been tanager-laden the previous day. This time, no birds. I wander about the clearing, saying good-bye to the jungle and looking for anything. I photograph a bullet ant and a stag horn beetle. Then, it is dark.

At dinner, I thank Megan and her group for enriching my stay. Megan replies that meeting people like me is one of the reasons she enjoys field work at Rara Avis. We acknowledge once again how it is different here. One needs to be independent and, as my experience proved, ready for almost anything.

Later, Megan and Emily invite me to look for some glass frogs—the frogs with transparent skins showing many of their internal organs. Although we don't find any, we do find a handsome Masked Treefrog. We are also present for exclamations from Giancarlo who is doing a night nature hike nearby with two guests from Switzerland. They had found a fer-de-lance in a small tributary that flows next to the hotel. It is always exciting to be near a creature so dangerous.

Back in my room, my bedding has been changed again. I stuff plastic bags in one hole and a towel in another. I feel sure that I am rat-proof. I have given Giancarlo what I hope is a generous tip, and he seems pleased. It has been a good day. I am reconciled; I can go home now.

CHAPTER **10**

I DON'T WANT TO GO!

Consume my heart away; sick with desire
And fastened to a dying animal
It knows not what it is;
— *William Butler Yeats*, "Sailing to Byzantium"

I am perfectly satisfied that I am being recycled, that only a few friends and family may remember me fondly, and that even this will vanish in a few short years. Nothing will remain. This is how nature works.
— *Gabor Levy*, "Farewell,"
July 1990 editorial, *American Laboratory*

I wake early, walk out naked on the porch, and raise my arms to a beautiful jungle sunrise. I am exhilarated, yet sad. The trip has been a dream, and it has gone so fast.

Now, we have gathered at the river. There are shouts. Giancarlo has spotted something. Megan is excited. She races into the shrubs and reeds and works her way down the small slope to the bank. There's something dark on a rock. A quick snatch; she has it. Running back, she displays her prize. It is an amphibian, coal black with a peculiar shape. The head is almost pointed as it narrows to a small triangular snout. It is a Black Narrowmouth Frog, also known as Costa Rica Nelson Frog (*Ctenophryne,* formerly *Nelsonophryne, aterrima*). Very difficult to see, I'm told, as if that information hasn't been conveyed by everyone's actions. Later, I read they are active mostly during nighttime rain and are otherwise hiding under leaf litter and other surface debris.

I take pictures. Megan wants to make sure she looks suitably excited and pleased in her photo. We are all happy. After views by everyone, Megan, true to the spirt of these researchers, scrambles back down the bank to release the frog where she caught it—hoping to ensure that its capture and inspection did not endanger its life. Now I feel forlorn. It is time to go.

Black Narrowmouth Frog

Where did the ten days go? There's still so much I want to see. I still haven't seen a White-vented Euphonia; it would be a new species for me. One of the researchers saw several yesterday. What about the Red-throated Caracara or the Barred Forest-falcon? I want more! Never satisfied.

I am doing what Buddhists describe as "clinging." Often, when times are good, we so desperately don't want the happy events to pass that we miss a satisfying present by holding on too tightly--trying too hard to prevent inevitable change. This time I recognize where my thinking has gone, and I tell myself, *Now is the time to indulge! Indulge in it. Feel all of it in every pore."* Fortunately, the older I've become, the easier it has become for me to wallow in the good times.

My clinging is mostly for health and freedom from pain. Of death itself I've lost my fear. Once, I came close. Hiking in the Grand Canyon in 2008, a rock broke under my feet triggering a landslide which propelled me over a cliff (*On Foot: Grand Canyon Backpacking Stories*, Vishnu Temple Press). This accident could have, maybe should have, taken my life. While lying in intensive care with multiple broken and shattered bones, I was overwhelmed by feelings of bad luck. To have died would have been much easier. I kept thinking, *I wouldn't have this pain. I wouldn't be looking at some loss of physical faculties. I wouldn't be looking at years of rehab and therapy and more surgery.* It would have been as if my life-switch had been abruptly turned off. Whatever transpired next, I wouldn't know. My life, as viewed by others, would have been described as "great" although "cut short." These were my thoughts as I coddled my misery.

Many have complimented me for my bravery for what I overcame. I see it as a desperate response to anxiety and frustration.

I performed physical exercise sometimes more than 40 hours in a week. Some of the movements were painful, but in my misery, I would not succumb to inactivity. Fortunately, exercise had always been my "go-to" method for calming my mind or deciding I was still ok no matter what had happened or what I'd done. Nearly every day, sometimes several times a day during my rehab, I'd think about quitting all the physical activity. Then, I would think, *What can I do, being what I am?* I recall nothing about striving for improvement, only that I strove for exhaustion.

I hated how the pain medications and sleep-inducing drugs felt in my mind. I quit taking them months before the doctors recommended, which increased my pain and probably my meanness. What I wanted was normalcy. By working out so hard, I was drained by evening. Without the drugs in my system, I could enjoy a glass of wine with my wife and then watch a movie with her. This would be followed by another night of "staring at the ceiling."

After each night of poor sleep, I'd wake up in a foul mood ready again to hate the world. My wife, who had retired from teaching, accepted a long-term substitute position for three months during, perhaps, the worst of this. We've said, not without gravity, that her working saved our marriage. Oh, we would have remained married, but I might have created an impassable barrier between us. She told me if I stayed so angry and difficult she wasn't going to let me drag her mental state down to where mine was. Her instinct for self-preservation was stronger than my anger and meanness. In a moment of clarity, I thanked her. I didn't promise to throttle my foul moods, because I couldn't—not then, but I felt relief when she announced that I couldn't hurt her anymore. Wallowing in misery doesn't seem like a good thing, but I did channel it into action. I exercised. Being exhausted at the end of the day, I wasn't great company, but at least by nightfall, I stopped being mean—until the next morning.

From there, things built. I improved. I had the immense great fortune of having come across a distinct, more advanced type of physical therapy (Institute of Physical Art) that was available a few hours away. Almost as important was our proximity to the Steadman Clinic an internationally renowned orthopedic surgery center, also a few hours away. Twice I was told in my hometown, "nothing can be done, take Ibuprofen" – only to be told at the Steadman Clinic, "We can do something about this." In the case of my right shoulder, the problem was 100% fixed by the surgery. Ten years later, I have full mobility and no pain. My left wrist couldn't be repaired but could be put into a state that was manageable. I had similar experiences with the physical therapy. Once my surgeons told me a severe ache in my lower back "...can't be fixed, take ibuprofen." I lamented the issue to my physical therapists. The guru of this field of study asked me to walk while he watched. "That's not it," he said, referring to the surgery center diagnosis. With a combination of therapy and manipulation, he and an associate eliminated my pain.

Thus, as I look back, I was more lucky than brave. I was lucky that exercise had always been my therapy when depressed. I was lucky my wife let me know she'd both stay by me and keep herself mentally safe. I was lucky that each day I could look forward to the glass of wine and a movie with her. Looking back, I can feel the sense of relief when she came home from work, and how by then I had dissipated the anger and frustration I'd been nourishing for the previous ten hours. I was also fortunate that my job as a consultant meant I worked at home, whenever I wanted to or needed to. I could put enough hours in by working a little every day, including weekends. I didn't have to manage going to an office where I wouldn't have been able to maintain so much exercise all through the day and be exhausted by early evening. Finally, I was lucky to have access to, and ability to afford, the medical care I received—certainly a cautionary tale these days.

I'm writing near the ten-year anniversary of that accident. I don't think it is possible to compare, for the most part, one part of life to another, but in the last ten years, there has been so much richness in my life, so many wonderful experiences. My life after the accident was worth far more than every moment of pain and anguish I endured. I'm so grateful I didn't die.

Now, I am sitting in our mountain cabin on an early spring morning. It is late March, mud season, but there is still considerable snow. A few birds are about, more than in December, but a fraction of what June will bring. I think of my family. I'd like to see my granddaughter or grandson sitting here at this age--contemplating a satisfying life. I won't see it. I'd be ~120 years old. I don't know when the lifeline between myself and them will be broken, but I know with certainty, it will be broken. That makes me sad, but I accept that reality.

I recently told my son, "I wish I'd live long enough to see how humans react to climate change." I'd like to see if there ever will be a recognition and vindication for those "who saw it coming" as author Terry Tempest Williams has said. Will more humans ever recognize their actual place in the cosmos? So many fascinating things have been learned during my lifetime. I'm curious about what's next. Curiosity may have killed the cat, but it's the best thing for a human.

Our lives are such a blink in time. I was reading to my grandson the other day. As with most young children, he's fascinated by dinosaurs. Two renderings that most of us can recognize are the Stegosaurus and the Tyrannosaurus. His plastic ones often play together. Yet, that is more historically inaccurate than picturing humans with a Tyrannosaurus. We know by rote, if not intuitively, how distant was their time: fully 65 million years ago is considered the end of the time of the dinosaur. The time of humans is closer to the time of the Tyrannosaurus Rex (68 to 66 million years ago) than is the Stegosaurus (151 to 156 million years ago). Now that's something. Humans have roamed the earth <2% as long as the dinosaurs, and that time period is much less than the distance of time between the death of the last Stegosaurus and the time of the Tyrannosaurs.

The sweep of evolution and life are so much greater than we are. Indeed, the marvel of life, and how and where it can appear and how and where it can thrive, is miraculous. No gods need be invoked; they only obstruct because they promise an "afterlife." I've come to believe that idea cheapens life on earth.

A seminal book for me, read when I was a practicing Christian, and, ironically, suggested by a minister friend, is entitled, *Something to Believe In* by Robert Short. The major premise is "there is no Hell." The book showed how focus on the afterlife keeps people from doing what they could to make everyone happier on this earth. Indeed, the central conclusion was "focusing on the afterlife makes people mean."

The author describes how focus on an afterlife caused many schisms in religious thought which led subsequently to conflict. Think of it this way: "If I am practicing 'the one true' religion correctly, I will go to heaven. If you are not, you will go to hell. Therefore, you are a source of evil in the world." Isn't this nothing but a system for producing enemies or at least for seeing those who are different as some threatening "other?" Clearly, this is the basis or justification for quite a lot of historical and current strife. I argue that eliminating aspiration for an after-life raises the value of the here and now. This is the life we and our fellow humans have now! This is it!

Wouldn't it be better if Islam, Christianity, and Judaism were not concerned with what came next, only with what is happening now? I was taught in Catholic elementary school that "there is a higher place in heaven for those who die for their faith." That kind of thinking was a root cause of the Crusades and the suicide bombings of 9-11 and its aftermath. Religions have often debased the idea of the "sanctity of life," by using belief in an afterlife as selling point for assembling an army and controlling the populace. I prefer how Skutch has put it in *Life Ascending*: "...hardly anything can so mitigate the grim prospect of [death] ...as the thought that, after we have gone, others who share our ideals will appreciate what we have appreciated, love what we have loved, care for what we have cared for, serve the causes that we can no longer advance. If we have given life to these others, or by our teaching or influence helped develop their minds and shape their ideals, and, above all, if in addition we love them personally, our departure will be less bitter to contemplate." Doesn't that beat being a martyr so you can have a "higher place in heaven," whatever that means?

Here again, I find myself in league with Buddhists. To a Buddhist, death is one of the "divine messengers." The message is that life is brief. Our imminent death is a reminder to pay attention to the present. Furthermore, I ascribe to Skutch's belief that "Reverent regard for the planet that bears us and steady determination to keep it fruitful and beautiful are our best hope for the unity of mankind." Here is recognition that all lives are precious, even the non-human ones. Reverent regard for the planet would change everything. We'd share. We'd cooperate. But we'll never do that—not most of us anyway. We don't get it. That's why I began the chapter with the quote from Yeats. Man "knows not what it is." We all, each of us, have a finite time on the planet. No matter how long we have, that time is incredibly brief. There is an excellent quote from one of the discredited books of Carlos Castaneda: "In a world where death is the hunter, my friend, there is no time for regrets or doubts. There is only time for decisions." Another way to put it is that there is no time for moods, but only time for action. And the desire for action comes from curiosity.

Curiosity is what leads to the joy the group at the river share examining the small, black, narrow-mouthed frog. After its release, those of us who have to leave start walking. We are an interesting group. Leading the way is a local youth who arrived with the tractor last night. I think he is supposed to be helping. The Swiss couple, who never interacted with the rest of us, is also leaving. The young man is striking because his trousers are red with vertical white stripes. Perhaps he's inspired by the colorful male manakins and hummingbirds. Then there are Emily and Eliot. Riding on Emily's chest, yesterday's travails a distant memory, is young Will, sleeping peacefully.

A fun part of this last morning is hearing a Laughing Falcon, and me knowing instantly what it is. Then I recognize the call of a White-fronted Parrot. I am learning! We are able to walk a good part of the way past Plástico until the departing tractor catches up with us and we have to ride.

We bounce along. I feel some envy for Emily and Eliot, on their way to four months of rainforest research in Panama, although I do wonder about the childcare. I am headed for a hospital in San José for a rabies shot, then on to a farewell dinner with Mario and Raquel.

While riding Eliot points out a soaring Black Hawk-Eagle—a bird I had not identified previously. We also see a couple of Long-tailed Tyrants—a mostly black flycatcher with a white head. The male is very distinctive, having central tail feathers that extend up to 13 centimeters beyond the rest of the tail. I am struck by Eliot and Emily's excitement at seeing the tyrants. These are common in Costa Rica, but I quickly learn that they are not found in Panama where Eliot and Emily have most of their tropical experience. How often one person's novelty is commonplace to another—which only shows how easily we forget that we are always in the presence of the wondrous.

Home begins to intrude in my thoughts. I am looking

forward to seeing Mary. Without her encouragement, this trip wouldn't have happened. Will there be another? Who can say? Suddenly, I'm filled with beautiful memories. I'm feeling not just the warmth of the weather, but the warmth of being part of this beautiful world. The enormity and complexity and beauty of it all causes me, once again, to appreciate Darwin's statement that "there is grandeur in this view of life..."

Darwin was describing the marvels of evolution and making his case against the concept of sudden creation. As I now use the word "view," and apply it to my own life, I know it has had its share of "rat bites." Many of the "bites" were self-inflicted. My intestines knot slightly as I write such a sentence. I have often been my worst enemy. Yet, I have recognized and reveled deeply in the presence of unimaginable beauty. I've felt the warmest and truest of emotions, especially the feeling of love. There has been much grandeur, and there is much grandeur yet. I don't want to leave.

ACKNOWLEDGEMENTS

I cannot give too much credit to my wife Mary. She has been the anchor in my life and my safe haven. Her love and support for me always made me feel OK, no matter how much the inner recesses of my mind were telling me something else. Her encouragement for the trip on which this book is based, and so many other activities I might not have undertaken, have made my life much happier and satisfying than it would have been.

Our daughter Ann and our son Adam give me hope for the future. Our exchanges regarding the world as seen through their eyes continue to influence me. They have patiently listened to many of my rants and have supported many of my causes and those of their own with fervor equal or greater than mine. They have always supported me with their love and concern, and both have chosen careers where they work for the future through education and therapy with youth. I'm so proud of them.

AND

With gratitude, I remember the people, animals, plants, the insects, creatures of the sky and sea, air and water, fire and earth, all whose joyful exertion blesses my life every day.

— *Jack Kornfield,* "Guided Meditation: Six essential Practices to Cultivate, Love, Awareness and Wisdom.

CPSIA information can be obtained
at www.ICGtesting.com
Printed in the USA
LVHW090351121220
673999LV00008B/154